# The Power of
# Logical Thinking

D0036805

Also by Marilyn vos Savant

*Ask Marilyn*

*More Marilyn*

*The World's Most Famous Math Problem:*
*The Proof of Fermat's Last Theorem and*
*Other Mathematical Mysteries*

*"I've Forgotten Everything I Learned in School!":*
*A Refresher Course to Help You Reclaim Your Education*

*Of Course I'm for Monogamy:*
*I'm Also for Everlasting Peace and an End to Taxes*

# The Power of Logical Thinking

Easy Lessons in the Art of
Reasoning . . . and Hard Facts
About Its Absence in Our Lives

## Marilyn vos Savant

St. Martin's Griffin ⚏ New York

For permission to reprint copyrighted material, grateful acknowledgment is made to the following sources:

*A. K. Dewdney, Ph.D.*: Excerpts from *200% of Nothing* by A. K. Dewdney. Copyright © 1993 by A. K. Dewdney, and "Mathematical Recreations" by A. K. Dewdney, from *Scientific American,* March. 1990 Copyright © 1990 by Scientific American, Inc. Reprinted by permission of the author.

*Ruma Falk, Ph.D.*: Excerpts from published and unpublished correspondence with the author. Reprinted by permission of Dr. Falk.

*Donald Granberg, Ph.D.*: "To Switch or Not to Switch," by Donald Granberg. Copyright © 1995 by Donald Granberg. Reprinted by permission of the author.

*W. W. Norton & Company, Inc.*: Excerpts from *How to Lie with Statistics* by Darrell Huff. Copyright © 1954 by Darrell Huff and Irving Geis. Copyright renewed 1982 by Darrell Huff and Irving Geis. Reprinted by permission of the publisher.

*Parade:* Excerpts from "Ask Marilyn" by Marilyn vos Savant. Copyright © 1990–1995 by Parade Publications, Inc. Reprinted by permission of Parade Publications, Inc.

*Scott Plous, Ph.D.*: Excerpts from *The Psychology of Judgment and Decision Making* by Scott Plous. Copyright © 1993 by Scott Plous. Reprinted by permission of the author.

*Design: Basha Zapatka*

Library of Congress Cataloging-in-Publication Data

Vos Savant, Marilyn
    The power of logical thinking : easy lessons in the art of reasoning—and hard facts about its absence in our lives / by Marilyn vos Savant.
        p.  cm.
    Includes bibliographical references.
    ISBN 0-312-15627-8
    1. Logic.  2. Logic—Problem, exercises, etc.  3. Critical thinking.  4. Reasoning
I. Title.
BC108.V67  1996
160—dc20                                                      95-40127
                                                                CIP

First St. Martin's Griffin Edition: May 1997

10  9  8  7  6  5  4  3  2  1

*Dedicated to*
*Martin Gardner*

*America's most beloved mathematician—and mine, too.*

# Contents

# Acknowledgments

▼

I thank Thomas McCormack, Chairman of St. Martin's Press, whose interest in numbers caused him to turn from more pressing matters and devote his time to a thoroughgoing mathematical examination of the first two sections of this book.

For the third section of this book, I gratefully acknowledge the invaluable contribution of Brian Doherty, formerly of the Cato Institute, who so successfully provided the research, ably assisted by Dean Stansel and Stephen Moore, also of the Cato Institute, Bruce Bartlett of the Alexis de Tocqueville Institute, and Christopher Frenze of the Joint Economic Committee of the United States Congress.

# Introduction

▼

I'm talking with a friend on the telephone. "If a drug test is 95 percent accurate, only the drug-users should be afraid of it, and I'm not a user," she says.

"But what about false positives?" I ask her.

"You mean lab error? I'll just take the test again."

"No, not lab error. A false positive is a result of the limitations of the test itself, and it can recur repeatedly."

"Well, we know the test isn't perfect, but 95 percent accuracy is high enough."

"Let's suppose the incidence of drug use in the population is 5 percent," I suggest. "We test someone at random, and he's positive. What do you think are the chances that he's a user?"

"95 percent, of course," she replies.

*"Not quite. They're only fifty-fifty."*

She pauses. "Would you repeat that?" (*See Part Two.*)

▼

I'm talking with another friend on the telephone. "So they're only offering me a $50,000 advance with the standard royalty rates and splits on the new book because they gave me $60,000 last time, and it didn't earn back the advance," he says.

"How much did it earn?" I ask him.

"$50,000. So the new offer makes sense."

"Did you have standard rates and splits last time?"

"No. It was my first book, you know."

"With standard rates and splits, what would it have earned?"

"Maybe $60,000," he replies.

*"So if they'd agreed to a worse deal for themselves on your last book, they'd be offering you a better deal now."*

He pauses. "Would you repeat that?" (*See Part One.*)

▼

Illogical thinking, especially with regard to numbers, is an educational condition that afflicts nearly every American, and even the literate lack a rigorous schooling in the art of reasoning, critical thinking, and problem-solving. This is a serious situation. Numbers are of paramount importance in our lives, now more than at any other time in history. Are we measuring up to that challenge? "The numbers" say "no." According to the Commission on Professionals in Science and Technology, 774 Americans received doctorates in mathematics in 1973. Although our population increased over the next twenty years, that number was down to 475 in 1993.

Moreover, as science progresses, direct human understanding becomes increasingly out of reach, and advanced mathematics provide the only tools for even the brightest people to manipulate highly complex principles. In short, if the math isn't perfect every step of the way, the conclusion can range from stupendous to silly, and we won't even know it. While giving cosmologists, astronomers, physicists, chemists, and mathematicians every bit of credit they deserve, what if just *one* of the seemingly countless steps on the way to a conclusion is counter-intuitive? Counter-intuitive problems, the solutions to which are strongly contrary to our intuitions, are everywhere, but the reason we don't notice that fact is because they're so . . . what else? . . . counter-intuitive. If we don't keep our reasoning skills honed, we draw mistaken conclusions daily, and unless we're nose-to-nose with evidence, we don't know it. (And sometimes not even then.) Here are a few examples. Do you find any of these statistics difficult to believe?

▼

• More people attended Paul Simon's 1991 concert in Central Park (upwards of 750,000) than the entire number of coalition forces in the Persian Gulf that year—including Americans, Saudis, British, Egyptian, and French (about 744,000). The number of people comprising what we considered a massive display of military manpower can spread out comfortably on the grass in Central Park with plenty of room to spare for things like dogs and lawn chairs and wine coolers. (And they already have.)

▼

• With the current world population of about 5½ billion, if we gathered everyone together temporarily and allotted each person a generous two foot-by-two foot patch of ground on which to stand, they'd cover an area of less than eight hundred square miles, only about the size of Jacksonville, Florida.

▼

• There are 3,871,000 miles of roadway in the country. If we take 50 feet as an arbitrary average width, including shoulders, this translates to 36,657 square miles of land used for automobile transportation. That's more than the entire state of Indiana. (And it doesn't include driveways, garages, and parking lots.)

▼

Part One of this book opens with an analysis of many provocative counter-intuitive problems that I've encountered in recent years. They illustrate how our own minds can work against us. All of the problems are confounding, but none is complex, and no sophisticated mathematics is needed to solve and understand them.

The Monty Hall Dilemma (it's not a paradox) is a good example. When it originally appeared in *Parade*, the response was so overwhelming that the magazine published an unprecedented four columns on the subject, twice devoting an entire page to just this one item. Then *The New York Times* wrote a front-page article about the furor, and when it went out over the news wires, front pages all over the country picked it up, inundating my clipping service with little drawings of goats, which were used to illustrate the story.

▼▼

Dear Marilyn:

Suppose you're on a game show, and you're given the choice of three doors. Behind one door is a car, the others, goats. You pick a door, say number 1, and the host, who knows what's behind the doors, opens another door, say number 3, which has a goat. He says to you, "Do you want to pick door number 2?" Is it to your advantage to switch your choice of doors?

Craig F. Whitaker
Columbia, Maryland

Dear Craig:

Yes, you should switch. The first door has a ⅓ chance of winning, but the second door has a ⅔ chance. Here's a good way to visualize what happened: Suppose there are a *million* doors, and you pick door number 1. Then the host, who knows what's behind the doors and will always avoid the one with the prize, opens them all except door number 777,777. You'd switch to that door pretty fast, wouldn't you?

Not long afterward, a London newspaper did an article about the problem and then called. It had drawn a record amount of mail, and would I consider writing for them every week? (Well, I couldn't do that, but I did write a couple more problems for them, and they teased their readers with them over a month of Sundays—literally—once doing a full page on the response.) And still later, I was at a meeting in Vienna and found myself in an eighteenth-century Austrian beer hall with about a hundred other people. My name was mentioned, and a fellow from Munich turned to me and said, "Oh, are you *that* Marilyn?!" He then proceeded to regale me with stories about how all three of the major German daily newspapers had written lengthy articles about the game-show problem, and it had even been re-created on national television—with live goats, yet. And when I returned to my apartment in Paris, there were telephone calls from half a dozen more countries.

That's where the episode began, actually. I had been sitting at the dining table in that same little old fourth-floor rental apartment on a quiet side street several years ago, puzzling over how to strengthen the seemingly innumerable weaknesses in my novel and straining to overhear the sounds of the choir from the church next door, when I heard the fax machine ring. It was a letter from Sara Brzowsky, my column's editor at *Parade*. "I'm afraid we have a problem," she wrote. "Mail is beginning to arrive from math professors everywhere—they say your answer in last Sunday's column is dead wrong."

Using a few of their letters, I wrote an additional brief column on the subject, assuming that a clarification would answer any outstanding questions. I was wrong. It brought the house down. Among the many thousands of letters from readers, we heard from prestigious universities all across the country in addition to the Department of Justice, the C.I.A., the F.B.I., and the Center for Defense Information, among many others, including pilots in the Persian Gulf. By the time the third column appeared in March of 1991, covering an entire page of *Parade* with letters of professional outrage, an additional explanation as detailed as I thought was reasonable, and a call for a national math experiment to test it, the storm had turned into a blizzard. Here's what the *Skeptical Inquirer* (Volume 15, Summer 1991) had to say about it:

> "When E. F. Hutton talks, people listen." Who can forget those indelible television commercials, in which the world came to a virtual standstill, focusing its sole attention upon that oracle of financial wisdom? . . . As a rule, when Marilyn vos Savant speaks in her weekly *Parade* magazine column, "Ask Marilyn," . . . people listen. In the instance discussed below, however, they may have been listening, but they weren't believing—although they should have been.

One reporter told me about his lunch with the head of the math department at a major university. The professor fumed for two full hours about my "gross stupidity" and "stubborn refusal" to acknowledge it, and the journalist came away genuinely confused—this was the best authority he could find. The next day, he received a desperate phone call from the shaken professor, pleading with him not to run the story. The professor had discovered he was wrong and feared that his students would never again believe a word he said. (The journalist ran the story, of course, but was kind enough to disguise the fellow.) And then there were the retractions—the charming ones and the grudging ones and one from a professor whose letter we'd published. He'd previously written on university letterhead, but then wrote back on plain white paper without even a return address. And we heard from the reader who'd sent the original question, who told us that strangers had found his number and were calling and pretending to be goats.

By the time we published the final column in July of 1991 we'd tallied the results of about a thousand real-life probability experiments, representing efforts by some fifty thousand students across the nation, from third grade through postgraduate school. Of course, the results of the experiment were determined by the parameters of the problem, but the satisfaction of that personal intellectual discovery turned out to be popular beyond what we'd anticipated and an effective learning experience, too.

Looking back over it all, what happened here? How do people handle counter-intuitive problems? What makes them so obstinate about changing their minds? A scholarly analysis of ten thousand of the Monty Hall Dilemma letters to "Ask Marilyn" was recently completed by Donald Granberg, Ph.D., at the University of Missouri Graduate School's Center for Research in Social Behavior. His paper, entitled "The Monty Hall Dilemma: To Switch or Not to Switch," is included in this book as an appendix.

The second half of Part One highlights some of the ways that illogical thinking manifests itself in financial matters. Which pay scale should we choose? Which price schedule? Which tollbooth? A reader once sent me the following quote from an ad for a jewelry store. "Up until the 1980's, we did 70 percent of our business with 30 percent of our customers," the owner declared. "Today, it's just about the opposite. Times have changed." They have?! Sounds like old times to *me*! And if these little examples can be so tricky, what must be happening with the big ones in our lives?

Part Two clarifies how numbers and statistics are misunderstood and used both to lead and mislead. If a person has an I.Q. of 200, is he or she "twice as smart" as a person with one of 100? Of course not. And we'll prove why. We'll also see—in addition to exploring facts about fallacies, fallacies about facts, and how our own minds work against us—that some of the most intriguing ramifications of the misuse of statistics are those afflicting the growing deployment of tests to detect drug use and human immunodeficiency virus type 1 (HIV-1) antibodies. The chances are excellent that you'll be tested at some point in your life, perhaps often, depending on your occupation and your insurance coverage. Can you answer the following question?

The Centers for Disease Control in Atlanta, Georgia, state that the two different well-known AIDS tests combined have greater than a 99 percent rate of accuracy, but only if they're taken repeatedly, over time and changing conditions. Furthermore, The Center states that .4 percent of Americans are "HIV positive." For the purpose of illustration, let's suppose that the accuracy is "only" 99 percent and that false positives account for just two-tenths of the 1 percent erroneous results. You're tested randomly, and the result is positive; you take the other test, and the result is positive, too. What are the chances that you're actually infected? Amazingly, the answer is only *fifty-fifty*.

Part Three illustrates how politicians exploit our innocence to affect our votes, using the 1992 presidential campaign as a superlative example. Never before have numerical issues been so much discussed, and seldom have numbers been so clearly used (and misused) as ammunition to hit political targets. Mathematics is simply logic with numbers, and when our ability to reason numerically is enhanced, so is our ability to reason in other ways. We gain power when we are able to think logically. Toward that end, we'll show examples of selective logic, specious reasoning, and outright sophistry, collected from the campaigns of Bill Clinton, George Bush, and Ross Perot. If Americans are better equipped to handle the numbers, we'll be less easily manipulated in future elections, and an unbiased study of Part Three can serve as a step toward the goal of heightening that critical consciousness.

Logical thinking empowers the mind in a way that no other kind of thinking can. It frees the highly educated from the habit of presuming every claim to be true until proven false. It enables average Americans to stand up against the forces of political correctness, see through the chicanery, and make independent decisions for themselves. And it is the bulwark against intellectual servitude for the underprivileged.

The best way to learn is by shattering our complacency, and this book takes aim at that target. For those of you readers to whom any sort of mathematics is anathema to your psychological stability, rest assured that our approach throughout will be far more logical than mathematical, requiring nothing more than uncommon sense and a little new-fashioned arithmetic. But prepare to be surprised more often than not. Whether in a real-life situation or in the form of a puzzle, many of these problems are sure to wreak havoc with anyone too content with his or her intuition.

And if you become convinced that I'm dead wrong about any answer, that I'm clearly a blockhead, and that you know the (obviously) correct answer, please do write to me at the address given in the back of the book. (I've taken the phone off the hook.) We're about to explore what promises to be controversial territory, and we invite you to come along for the ride.

# Part One

HOW OUR OWN MINDS CAN
WORK AGAINST US

# One

## Lessons Set in Everyday Life

*The war was started as the result of a mistaken intuitive "calculation" which transcended mathematics. We believed with a blind fervor that we could triumph over scientific weapons and tactics by means of our mystic will . . . . The characteristic reliance on intuition by Japanese had blocked the objective cognition of the modern world.*

Nyozekan Hasegawa
*The Lost Japan*, 1952

Although we shouldn't underestimate the damage caused by reliance on intuition, our personal experience with it often takes a more prosaic form. Not long ago, a New Jersey attorney asked that I submit a certification on an issue of probability. (In an action before the Union County Superior Court, the judge had suggested that the attorney contact me in order to calculate the chances of being among the winners in a lottery that had an impact on her client.) The problem was presented as follows:

What are the chances of any one person winning any one prize in a lottery where there are twenty prizes of equal value to be awarded one each to twenty different winners in a pool of thirty-three contestants, each of whom is represented (by a name, number, ticket, etc.) only once? (All prizes will be awarded, none will be shared, all chances have been distributed.) With the same conditions, how does this compare to a drawing wherein thirteen people are vying for only one prize?

The attorney then asked me one more question: "What adjustments would have to be made in order to equalize the odds of a contestant's prevailing in the second lottery with those odds existing in the first lottery?"

Here are the relevant facts: The City of Elizabeth was ordered by the Superior Court to conduct a lottery for the award of taxicab medallions. In the first lottery, thirty-two drivers (according to a certain definition) participated, but another person was disallowed. Later, however, a second lottery was held in order to allow that person an equal chance, after all. In this "mini-lottery," thirteen contestants participated—the unsuccessful twelve from the first pool plus the new person—and one additional taxicab medallion was awarded.

Was this fair?

Our intuition says, "Yes," but the objective answer is, "No." Here's why: In the first lottery, twenty prizes were awarded to individuals out of a total pool of thirty-two contestants. This means that the chances of any one individual winning a prize were 20/32 (or .625). (This translates to odds of 20:12—twenty people will win; twelve people will lose.)

In the second lottery, one prize was awarded to an individual out of a total pool of thirteen contestants. This means that the chance of any one individual winning the one prize was 1/13 (or .076923). (This translates to odds of 1:12—one person will win; twelve people will lose.)

In short, the chances of any one individual winning a prize in the first lottery were slightly more than eight times greater than that individual's chance of winning a prize in the second lottery. (And if there had been thirty-three contestants in the first lottery, the chances of any one individual winning a prize were slightly less than eight times greater than that individual's chance of winning a prize in the second lottery.) In other words, the first lottery was about eight times easier to win than the second lottery. The two lotteries were greatly unequal.

But after the first lottery was already held, was there an easy solution to the problem?

Yes. Assuming that a thirty-third individual had not been allowed to take part in the first lottery, and a second lottery was going to be held in order to allow that individual an equal chance at one more prize, we could just write the new contestant's name on twenty slips of paper and write no name at all on twelve more slips. (This gives him/her exactly the same chance the other contestants had: 20/32.) A slip of paper is drawn. If his name is on it, he wins the prize. If not, no prize is awarded. (This latter scenario illustrates the degree to which the thirty-third individual was disadvantaged at the actual second lottery.)

The judge asked for further explanation, which wasn't surprising; even the most fair and cautious of us have trouble disregarding our intuitions. We might have reasoned that after the first nineteen prizes (out of twenty) were awarded to nineteen people (out of thirty-two), the remaining thirteen people had a 1/13 chance of winning the one last prize.

And in a narrow sense, that's true. But only if the lottery were fresh at that point. That's because these people started out with a "value" (in chances) of 20/32 (.625), and over the course of the lottery, "spent" their chances (throughout the twenty drawings) down to 1/13 (.076923). That's a large expenditure.

So the last prize is awarded, and we're left with twelve people. Now we add one new prize and one new individual. This is a fresh lottery, and that new individual does indeed have a chance of 1/13 (.076923)—the same as the others. But he/she didn't get the opportunity to "spend" a greatly larger chance in the first lottery, the way the others did.

Let's say you're given these two options:

1. You can take part in a lottery where you have a 20/32 chance of winning *and,* if you should be unlucky enough to lose, have an additional 1/13 chance of winning in a new lottery.

or

2. You can take part only in a lottery where you have a 1/13 chance of winning.

Of course, we'd all choose option number 1, which offers far more opportunity for winning than number 2. (Number 1 was the position allowed all the other people, but number 2 was the only position allowed the "new" individual. That's why he/she was at a great disadvantage.)

## The Monty Hall Dilemma

One of the best examples of the counter-intuitive problems, and now the most famous, is known as the Monty Hall Paradox, but I call it the Monty Hall Dilemma because it isn't a paradox at all. The problem itself is not difficult to grasp, but because the intuitive answer seems so obvious—and that "obvious" answer is dead wrong—it ensnares many people. When a problem appears complex—or even just a little out of the ordinary, like the taxicab medallion problem—most people don't become utterly convinced that an answer other than the one given is correct. They don't take an early position; they don't take a position at all. But if a problem appears simple, another phenomenon occurs. That's what happened when *Parade* published the following in the "Ask Marilyn" column.

Dear Marilyn:

Suppose you're on a game show, and you're given the choice of three doors. Behind one door is a car, behind the others, goats. You pick a door, say number 1, and the host, who knows what's behind the doors, opens another door, say number 3, which has a goat. He says to you, "Do you want to pick door number 2?" Is it to your advantage to switch your choice of doors?

Craig F. Whitaker
Columbia, Maryland

Dear Craig:

Yes, you should switch. The first door has a 1/3 chance of winning, but the second door has a 2/3 chance. Here's a good way to visualize what happened: Suppose there are a *million* doors, and you pick door number 1. Then the host, who knows what's behind the doors and will always avoid the one with the prize, opens them all except door number 777,777. You'd switch to that door pretty fast, wouldn't you?

Well, I thought the explanation was clear and had no idea that anyone would take issue with it, so it was with great surprise that I found myself forced to publish the following column.

Dear Marilyn:

Since you seem to enjoy coming straight to the point, I'll do the same. In the following question and answer, you blew it! Let me explain. If one door is shown to be a loser, that information changes the probability of either remaining choice, *neither of which has any reason to be more likely,* to 1/2. As a professional mathematician, I'm very concerned with the general public's lack of mathematical skills. Please help by confessing your error and in the future being more careful.

Robert Sachs, Ph.D.

George Mason University

You blew it, and you blew it big! Since you seem to have difficulty grasping the basic principle at work here, I'll explain. After the host reveals a goat, you now have a one-in-two chance of being correct. Whether you change your selection or not, the chances are the same. There is enough mathematical illiteracy in this country, and we don't need the world's highest I.Q. propagating more. Shame!

S. S., Ph.D.

University of Florida

Your answer to the question is in error. But if it is any consolation, many of my academic colleagues have also been stumped by this problem.

Barry Pasternack, Ph.D.

California Faculty Association

Dear Readers:

Good heavens! With so much learned opposition, I'll bet this one is going to keep math classes all over the country busy on Monday.

My original answer is correct. But first, let me explain why your answer is wrong. The winning chances of 1/3 on the first choice can't go up to 1/2 just because the host opens a losing door. To illustrate this, let's say we play a shell game. You look away, and I put a pea under one of three shells. Then I ask you to put your finger on a shell. The chances that your choice contains a pea are 1/3, agreed? Then I simply lift up an empty shell from the remaining other two. As I can (and will) do this regardless of what you've chosen, we've learned nothing to allow us to revise the chances on the shell under your finger.

The benefits of switching are readily proven by playing through the six games that exhaust all the possibilities. For the first three games, you choose number 1 and "switch" each time, for the second three games, you choose number 1 and "stay" each time, and the host always opens a loser. Here are the results.

|         | DOOR 1 | DOOR 2 | DOOR 3 |                    |
|---------|--------|--------|--------|--------------------|
| GAME 1  | AUTO   | GOAT   | GOAT   | Switch and you lose. |
| GAME 2  | GOAT   | AUTO   | GOAT   | Switch and you win.  |
| GAME 3  | GOAT   | GOAT   | AUTO   | Switch and you win.  |
|         | DOOR 1 | DOOR 2 | DOOR 3 |                    |
| GAME 4  | AUTO   | GOAT   | GOAT   | Stay and you win.  |
| GAME 5  | GOAT   | AUTO   | GOAT   | Stay and you lose. |
| GAME 6  | GOAT   | GOAT   | AUTO   | Stay and you lose. |

When you switch, you win 2/3 of the time and lose 1/3, but when you don't switch, you only win 1/3 of the time and lose 2/3. You can try it yourself and see.

Alternatively, you can actually play the game with another person acting as the host with three playing cards—two jokers for the goat and an ace for the prize. However, doing this a few hundred times to get statistically valid results can get a little tedious, so perhaps you can assign it as extra credit—or for punishment! (*That'll* get their goats!)

I was convinced this would end the matter. But, oddly, it only seemed to fan the flames, even though the explanation is straightforward, not at all misleading or confusing, and my critics included many distinguished people. However, I decided to persist in my efforts to be understood because I didn't want to let so many people continue to believe that I was wrong and that I simply wouldn't admit it. (I don't know if this is a negative or positive facet of my personality: I don't especially mind being wrong, but I do particularly mind being thought wrong when I'm right.) Fortunately, this time I thought of a way to prove the answer to everyone's satisfaction.

▼▼

Dear Marilyn:

You're in error, but Albert Einstein earned a dearer place in the hearts of people after he admitted his errors.

Frank Rose, Ph.D.
University of Michigan

I have been a faithful reader of your column, and I have not, until now, had any reason to doubt you. However, in this matter (for which I do have expertise), your answer is clearly at odds with the truth.

James Rauff, Ph.D.
Millikin University

May I suggest that you obtain and refer to a standard textbook on probability before you try to answer a question of this type again?

Charles Reid, Ph.D.
University of Florida

I am sure you will receive many letters on this topic from high school and college students. Perhaps you should keep a few addresses for help with future columns.

W. Robert Smith, Ph.D.
Georgia State University

You are utterly incorrect about the game-show question, and I hope this controversy will call some public attention to the serious national crisis in mathematical education. If you can admit your error, you will have contributed constructively towards the solution of a deplorable situation. How many irate mathematicians are needed to get you to change your mind?

E. Ray Bobo, Ph.D.
Georgetown University

I am in shock that after being corrected by at least three mathematicians, you still do not see your mistake.

Kent Ford
Dickinson State University

Maybe women look at math problems differently than men.

Don Edwards
Sunriver, Oregon

You are the goat!

Glenn Calkins
Western State College

You made a mistake, but look at the positive side. If all those Ph.D.'s were wrong, the country would be in some very serious trouble.

Everett Harman, Ph.D.
U.S. Army Research Institute

Dear Readers:

Gasp! If this controversy continues, even the *postman* won't be able to fit into the mailroom. I'm receiving thousands of letters, nearly all insisting that I'm wrong, including the Deputy Director of the Center for Defense Information and a Research Mathematical Statistician from the National Institutes of Health! Of the letters from the general public, 92 percent are against my answer, and of the letters from universities, 65 percent are against my answer. Overall, nine out of ten readers completely disagree with my reply. But math answers aren't determined by votes.

We're now receiving far *more* mail, and even newspaper columnists are joining in the fray! The day after the second column appeared, lights started flashing here at the magazine. Telephone calls poured into the switchboard, fax machines churned out copy, and the mailroom began to sink under its own weight. Incredulous at the response, we read wild accusations of intellectual irresponsibility, and, as the days went by, we were even more incredulous to read embarrassed retractions from some of those same people!

So let's look at it again, remembering that the original answer defines certain conditions, the most significant of which is that *the host always opens a losing door on purpose.* (There's no way he can always open a losing door by chance!) Anything else is a different question.

The original answer is still correct, and the key to it lies in the question, "*Should you switch?*" Suppose we pause at that point, and a UFO settles down onto the stage. A little green woman emerges, and the host asks her to point to one of the two unopened doors. The chances that *she'll* randomly choose the one with the prize are 1/2, all right. But that's because she lacks the advantage the *original* contestant had—the help of the host. (Try to forget any particular television show.)

When you first choose door number 1 from three, there's a 1/3 chance that the prize is behind that one and a 2/3 chance that it's behind one of the others. *But then the host steps in and gives you a clue.* If the prize is behind number 2, the host shows you number 3, and if the prize is behind number 3, the host shows you number 2. So when you switch, you win if the prize is behind number 2 *or* number 3. *You win either way!* But if you *don't* switch, you win only if the prize is behind door number 1.

And as this problem is of such intense interest, I'll put my thinking to the test with a nationwide experiment. This is a call to math classes all across the country. Set up a probability trial exactly as outlined below and send me a chart of all the games along with a cover letter repeating just how you did it, so we can make sure the methods are consistent.

One student plays the contestant, and another, the host. Label three paper cups number 1, number 2, and number 3. While the contestant looks away, the host randomly hides a penny under a cup by throwing a die until a one, two, or three comes up. Next, the contestant randomly points to a cup by throwing a die the same way. Then the host purposely lifts up a losing cup from the two unchosen. Lastly, the contestant "stays" and lifts up his original cup to see if it covers the penny. Play "not switching" two hundred times and keep track of how often the contestant wins.

Then test the other strategy. Play the game the same way until the last instruction, at which point the contestant instead "switches" and lifts up the cup *not* chosen by anyone to see if it covers the penny. Play "switching" two hundred times, also.

And here's one last letter.

Dear Marilyn:
You are indeed correct. My colleagues at work had a ball with this problem, and I dare say that most of them, including me at first, thought you were wrong!

Seth Kalson, Ph.D.
Massachusetts Institute of Technology

Dear Dr. Kalson:
Thanks, M.I.T. I needed that!

Personally, I regarded the probability trial as more of a proof than an experiment, but it worked. Here's the fourth (and final!) column *Parade* ran on the subject. (And note that the credentials of my supporters are somewhat different from those of my detractors!)

▼▼

Dear Marilyn:

In a recent column, you called on math classes around the country to perform an experiment that would confirm your response to a game show problem. My eighth-grade classes tried it, and I don't really understand how to set up an equation for your theory, but it definitely does work! You'll have to help rewrite the chapters on probability.

Pat Gross, Ascension School
Chesterfield, Missouri


Our class, with unbridled enthusiasm, is proud to announce that our data support your position. Thank you so much for your faith in America's educators to solve this.

Jackie Charles,
Henry Grady Elementary
Tampa, Florida


My class had a great time watching your theory come to life. I wish you could have been here to witness it.

Pat Pascoli, Park View School
Wheeling, West Virginia


Seven groups worked on the probability problem. The numbers were impressive, and the students were astounded.

R. Burrichter,
Webster Elementary School
St. Paul, Minnesota


You could hear the kids gasp one at a time, "Oh my gosh. She was right!"

Jane Griffith, Magnolia School
Oakdale, California


I must admit I doubted you until my fifth-grade math class proved you right. All I can say is WOW!

John Witt, Westside Elementary
River Falls, Wisconsin

Thanks for that fun math problem. I really enjoyed it. It got me out of fractions for two days! Have any more?

Andrew Malinoski,
Mabelle Avery School
Somers, Connecticut

You have taken over our Mathematics and Science Departments! We received a grant to establish a Multimedia Demonstration Project using state-of-the-art technology, and we set up a hypermedia laboratory network of computers, scanners, a CD-ROM player, laser disk players, monitors, and VCRs. Your problem was presented to 240 students, who were introduced to it by their science teachers. They then established the experimental design while the mathematics teachers covered the area of probability. Most students and teachers initially disagreed with you, but during practice of the procedure, all began to see that the group that switched won more often. We intend to make this activity a permanent fixture in our curriculum.

Anthony Tamalonis, Arthur S. Somers
Intermediate School 252
Brooklyn, New York

I put my solution of the problem on the bulletin board in the physics department office at the Naval Academy, following it with a declaration that you were right. All morning I took a lot of criticism and abuse from my colleagues, but by late in the afternoon most of them came around. I even won a free dinner from one overconfident professor.

Eugene Mosca, Ph.D.,
U.S. Naval Academy
Annapolis, Maryland

After considerable discussion and vacillation here at the Los Alamos National Laboratory, two of my colleagues independently programmed the problem, and in 1 million trials, switching paid off 66.7 percent of the time. The total running time on the computer was less than one second.

> G.P. DeVault, Ph.D.,
>     Los Alamos National Laboratory
> Los Alamos, New Mexico

Now 'fess up. Did you really figure all this out, or did you get help from a mathematician?

> Lawrence Bryan
> San Jose, California

Dear Readers:

Wow! What a response we received! It's still coming in, but so many of you are so anxious to hear the results that we'll stop tallying for a moment and take stock of the situation so far. We've received thousands of letters, and of the people who performed the experiment by hand as described, the results are close to unanimous: You win twice as often when you change doors. Nearly 100 percent of those readers now believe it pays to switch.

But many people tried performing similar experiments on computers, fearlessly programming them in hundreds of different ways. Not surprisingly, they fared a little less well. Even so, about 97 percent of them now believe it pays to switch.

And plenty of people who *didn't* perform the experiment wrote, too. Of the general public, about 56 percent now believe you should switch compared with only 8 percent before. And from academic institutions, about 71 percent now believe you should switch compared with only 35 percent before. (Many of them wrote to express utter amazement at the whole state of affairs, commenting that it altered their thinking dramatically, especially about the state of mathematical education in this country.) And a very small percentage of readers feel convinced that the furor is resulting from people not realizing that the host is opening a losing door on purpose. (But they haven't read my mail! The great majority of people understand the conditions perfectly.)

And so we've made progress! Half of the readers whose letters were published in the previous columns have written to say they've changed their minds, and only this next one of them wrote to state that his position hadn't changed at all.

Dear Marilyn:

I still think you're wrong. There is such a thing as female logic.

Don Edwards
Sunriver, Oregon

Dear Don:

Oh hush, now.



Massimo Piattelli-Palmarini stated in *Bostonia* (July/August 1991) that, ". . . no other statistical puzzle comes so close to fooling all the people all the time. . . . The phenomenon is particularly interesting precisely because of its specificity, its reproducibility, and its immunity to higher education."

He goes on, "Think about it. Ask your brightest friends. Do not tell them, though (or at least not yet), that even Nobel physicists systematically give the wrong answer, and that they *insist* on it, and are ready to berate in print those who propose the right answer."

As I replied to the readers of the *Skeptical Inquirer* (Volume 16, Winter 1992), ". . . virtually all of my critics understood the intended scenario. I personally read nearly three thousand letters (out of the many additional thousands that arrived) and found nearly every one insisting simply that because two options remained (or an equivalent error), the chances were even. Very few raised questions about ambiguity, and the letters actually published in the column were not among those few.

"But for those readers now interested more in the analysis of ambiguity, let me offer the following notes. When I read the original question as it was sent by my reader, I felt it didn't emphasize enough that the host always opens a door with a goat behind it, so I added that to the answer to make sure everyone understood. And as for whether the host offers the switch each time, I don't see that as a valid objection. It wasn't offered as a factor, so the original is the paradigm. The contestant chooses a door each time; the host opens a door each time. (The contestant doesn't choose a door and *open* it next time; the host doesn't open the *contestant's* door next time; the host doesn't offer the contestant *money* next time.)

"Just because a similar game show appeared on television doesn't warrant the assumption that the published problem involves considerations as subjective as creating audience excitement or saving the sponsor money!" Just as assumptions stated as conditions in the original question and answer cannot be changed after the fact (!), neither can new ones be added.

The foregoing problem might seem more relevant to game-show contestants and street-corner swindlers, but that's because problems like it are usually set in Dick-and-Jane circumstances rather than Boris-and-Mikhail ones, presumably because people find learning easier in familiar situations. But Boris-and-Mikhail are no less relevant, and Dick-and-Jane are easily the more artificial. We do indeed apply the math we learned when we were twenty-one (and sitting in a classroom) when we're forty-one (and sitting in a war room), and the ramifications of that fact are underestimated at our peril. Consider, for example, the broad implications of misunderstanding the consequences of choices made on the battlefield, in the air, or on the high seas. Logical thinking can be a matter of life and death—your own or many more.

### The Psychological Component

Amos Tversky and Daniel Kahneman pose such a a life-and-death problem in *Judgment Under Uncertainty,* which this reader sent to me paraphrased:

▼▼

Dear Marilyn:

My psychology class received the following question: Threatened by a superior enemy force, the general faces a dilemma. His intelligence officers say his soldiers will be caught in an ambush in which all six hundred of them will die unless he leads them to safety by one of two available routes. If he takes the first, four hundred will die. If he takes the second, there's a one-third chance that none will die, and a two-thirds chance that six hundred will die. Which route should he take?

The answer we were given was that it doesn't matter, because the chances are the same in either case. The wording is supposed to trick you into thinking there's a difference. But many of us in the class failed to follow this line of reasoning. Is there a mathematical solution to this?

J. Cooke
Honolulu, Hawaii

John Allen Paulos wrote about the same problem in his book *Innumeracy,* noting that, "Most people (four out of five) faced with this choice opt for the second route, reasoning that the first route will lead to four hundred deaths, while there's at least a probability of 1/3 that everyone will get out okay if they go for the second route."

This was my own analysis:

Dear Reader:

If we were talking about civilians, and if the intention of the general were simply to save as many lives as possible, and if this choice were to be repeated again and again, then no, there would be no difference mathematically. But you received this question in psychology class, not math class, and armies are not created to save their own lives. They have other objectives, often requiring significant sacrifice of life. For example, if only the survival of two hundred soldiers is required to secure a particular location and win the war, the general should take the first route. But if three hundred are required, he must risk the second.

Now let's suppose you're that same general surrounded by that same enemy, but this time your intelligence officers inform you that if you choose the first route, two hundred soldiers will live, whereas if you choose the second route, there's a one-third chance that all six hundred will live, and a two-thirds chance that all six hundred will die. Which route do you take *now*?

According to Paulos, "Most people (75 percent) choose the first route, since two hundred lives can definitely be saved that way, whereas the probability is 2/3 that the second route will result in even more deaths." But the two questions are just two ways of posing the identical problem and the fact that most people will change their decision based on the way it is worded highlights the evidence that our intuition is inadequate.

### Probability Problems

This next problem, which is related to the Monty Hall Dilemma, causes similar controversy. Even people who came to agree with my game-show answer wrote back to insist that this one was wrong.

Dear Marilyn:

Three prisoners on death row are told that one of them has been chosen at random for execution the following morning, but the other two are to be freed. One privately begs the warden to at least tell him the name of one other prisoner who will be freed, and the warden relents. "Susie will go free," he says. Suddenly horrified, the first prisoner says that because he is now one of only two remaining prisoners at risk, his chances of execution have risen from one-third to one-half! What should the warden do?

Marvin M. Kilgo III
Camden, South Carolina

Dear Marvin:

Going home and not coming back until the following afternoon sounds like a pretty good idea. Even though there are only two remaining prisoners at risk, the first prisoner still has only a one-third chance of execution. Oddly enough, however, things don't look so good for the other one, whose chances have now gone up to two-thirds!

We decided for editorial reasons not to publish the following in the column, but I think a little more explanation is probably still in order, so here it is now:

▼▼

Dear Marilyn:

You have done it again! I must strongly recommend that you either get a new reference or take a course (beginning college level) in probability. The prisoner's chances for execution did indeed go from one-third to one-half. People believe what you say, so give them correct information.

Stephen Van Fossan, Ph.D.
San Diego, California

You cannot correctly apply feminine logic to odds. The new situation is one of two equal chances.

Charlie Paine
Richmond, Virginia

Dear Readers:

There's no such thing as feminine logic, and the original answer is correct. The questioning prisoner had a one-third chance of execution, and we've learned nothing to change that; however, we *have* learned something about the two others. Because they'll include the doomed prisoner two-thirds of the time, and because the warden will never name that one, the prisoner who isn't named has that two-thirds chance of execution.

I wouldn't want my readers to grow complacent, so we published this next question and answer. Unfortunately, however, there was no room for an explanation, a mistake that I don't intend to make again!

▾▾

Dear Marilyn:

A shopkeeper says she has two new baby beagles to show you, but she doesn't know whether they're male, female, or a pair. You tell her that you want only a male, and she telephones the fellow who's giving them a bath. "Is at least one a male?" she asks him. "Yes!" she informs you with a smile. What is the probability that the *other* one is a male?

> Stephen I. Geller
> Pasadena, California

Dear Stephen:

One out of three.

Mail began to arrive in response, which I somehow managed to find surprising. It was beginning to seem that understanding one counter-intuitive problem didn't necessarily mean that a person would understand the next one to come along. Or maybe these were just different readers. Anyway, this next column was the, perhaps predictable, result.

▾▾

Dear Marilyn:

This is regarding the problem where a shopkeeper has two new baby beagles, but she doesn't know yet whether they're male, female, or a pair. There are only three possible explanations for your answer.

1. You are considering the possibility that the second puppy has been neutered, but because you did not make this clear, you leave open the next explanation.
2. You are (still) confused about the difference between game theory and probability. If so, I suggest you brush up on this information.

3. Your incorrect responses are intentional, done as a means
   to solicit mail. If so, there has to be a better way.
                                        James Larsen, Ph.D.
                                        Wright State University


I disagree with your answer. The observer knows the first dog is a male.
The odds that the other is a male are fifty-fifty.
                                        Richard Jones, Ph.D.
                                        University of New Haven


I believe you're wrong! Before you know that one of the baby beagles
is male, the chances are one out of three. But once the above fact is known,
the odds change to fifty-fifty.
                                        Edward Weiss, Ph.D.
                                        Bethel College


Okay. But here's what puzzling to me. If there were only one beagle
pup, the probability of it being male would be one out of two. What can
explain why the presence of another beagle pup should affect the probability that the pup in question is a male? I know this is cute, but how far
away does the second beagle have to be in order for it not to affect the
sex of the first beagle?
                                        Steve Marx
                                        Worcester, Massachusetts


Dear Readers:
    The original answer is correct. We didn't define a "first beagle" or a
"second beagle," and so either beagle can be in a doghouse on the moon,
and it would still affect the outcome.
    If we could shake a pair of puppies out of a cup the way we do dice,
there are four ways they could land: male/female, or female/male, or male/
male, or female/female. So there are three ways in which at least one of
them could be a male. And as the partner of a male in those three is a
female, a female, or another male, the chances of that partner being a male
are only one out of three.
    The key is that we didn't specify *which* beagle was a male, so it can be
either one. If we'd said instead, "The one nibbling your ankle is a male;

what are the chances that the one sleeping is a male?" the odds would be fifty-fifty. But we just specified that "at least one" was a male, so we don't know which it is and whether its partner is awake or asleep. And this means that the chances of that partner being a male are only one out of three.

Again I was surprised when the mail didn't stop, but those days of innocence are long gone. We decided not to publish the following correspondence (in order to keep the column from turning into a logic/math/probability course), but I couldn't resist including it in this book. (Especially the last letter, from Dr. Dowling.)

▼▼

Dear Marilyn:

This is regarding the problem where a shopkeeper has two new baby beagles but doesn't yet know their sex. In a second column on the subject, you stated that your answer was correct and said that "the key is that we didn't specify *which* beagle was a male, so it can be either one."

You are confused about probability. It makes no difference whether you specify which of the two puppies you are talking about; what is crucial is that you are talking about only these two and no others. Exactly the same sort of problems arise in genetic counseling, where the consequences of flawed understanding are very serious.

Steven M. Carr, Ph.D.
Memorial University of Newfoundland

From a geneticist, you unnecessarily attempted to determine probabilities for an individual based on the probabilities of pairwise combinations of two individuals. More parsimonious explanations came from your other readers: The 50 percent probability of the sex of one pup does not influence that of the other.

H. Glenn Hall, Ph.D.
University of Florida

I'm afraid that you are totally, irrevocably and unequivocally, incorrect in your answer.

Stephen D. Wolpe, Ph.D.
Genetics Institute

The fact is that knowing the sex of either puppy leaves only two possible combinations out of the original four possible combinations. To add a modicum of believability to my analysis, I am an analyst for a supercomputing company.

Bruce Greer
Hillsboro, Oregon

You are clearly right. I think the reason so many people have disputed your answer is because they are not math teachers. I've been teaching math all my life, and I really get a kick out of your problems.

Thomas E. Hurst
Chicopee High School

Dear Readers:

The original answer is correct. We're receiving quite a bit of mail, and most of the critics say a version of the following: "If there's only one puppy left, it's either a male or a female, and therefore the chances are fifty-fifty that it's a male." But that's as flawed as saying that tomorrow morning, the sun is either going to rise or it isn't, and therefore the chances are only fifty-fifty that the sun is going to rise tomorrow.

Here's another way to look at it. The fellow who's bathing the puppies acknowledges, through the shopkeeper, that at least one is a male. That means he's talking about a pair of puppies, not just one. And asking about the probability that the other one is a male is a question about that pair, not a specific dog. In other words, it's the same as asking whether *both* are male.

Here's one last letter, and I think we should all be very thankful that this man got it right!

Dear Marilyn:

You are correct again, and it is all those cocky fellows with an attitude who need a refresher course in math.

Jonathan P. Dowling, Ph.D.
Department of the Army
United States Army Missile Command
Redstone Arsenal, Alabama

This next reader actually wrote a very sober letter about a hat containing three playing cards, but I wanted to change them to pancakes instead, so we phoned him to ask his permission. He readily gave it, although I wonder what he thought of us. After all, . . . pancakes? I just wanted to see if anyone would write and point out the obvious—that no one puts pancakes in hats. (Although I suppose that relatively few people put cards in hats, either!)

▼▼

Dear Marilyn:

You have a hat in which there are three pancakes: One is golden on both sides, one is brown on both sides, and one is golden on one side and brown on the other. You withdraw one pancake, look at one side, and see that it is brown. What is the probability that the other side is brown?

Robert H. Batts
Acton, Massachusetts

Dear Robert:

It's two out of three. The pancake you withdrew had to be one of only two of them: the brown/golden one or the brown/brown one. And of the three brown sides you could be seeing, two of them also have brown on the other side.

Well, no one wrote about the pancakes, but quite a few wrote about the answer.

▼▼

Dear Marilyn:

I look forward every week to your column, but I disagree with your analysis of the "three pancakes in a hat" question. I believe the odds are fifty-fifty. Since we know that we hold one of two, and the other sides are either brown or gold, it is equally likely that either color shows up when we flip our pancake over. Don't you agree? (I waited to see if you announced a flood of mail on this one!)

Elmer Mooring, Jr.
Johns Hopkins University

Your answer about the pancakes is wrong. It should have been one chance in two (50 percent). Only two pancakes have brown sides, and one of them has brown on only one side. There is a 50 percent chance that you are looking at the one with brown on both sides. I have been told that you have never publicly admitted to being wrong. Is this true?

R. Larry Marchinton, Ph.D.
University of Georgia

Dear Readers:

Don't worry. Whenever I'm wrong, we announce it in this column loud and clear. Some readers are still certain that earlier probability answers are wrong and that I simply won't admit an error. But that's not the case, and this one wasn't an exception. The original answer is correct. Note that it's easier to discover a brown side on a brown/brown pancake than on a brown/gold pancake.

Here's another way to look at it. Before you pull out any pancake at all, what are the chances that you'll pull out a pancake with sides that match? They're two out of three right? So if you pull out a pancake and see a gold side, the chances that the other side is also gold is 2/3. Likewise, if you pull out a pancake and see a brown side, the chances that the other side is also brown is 2/3.

I was pleased to notice the incomparable Martin Gardner's name on the following letter, which turned out to be the start of a fine friendship.

▼▼

Dear Marilyn:

The big flap over your game-show probability problem prompts me to send another: The Greens and Blacks are playing bridge. After a deal, Mr. Brown, an onlooker, asks Mrs. Black: "Do you have an ace in your hand?" She nods. There is a certain probability that her hand holds at least one other ace.

After the next deal, he asks her: "Do you have the ace of spades?" She nods. Again, there is a certain probability that her hand holds at least one other ace. Which probability is greater? Or are they both the same?

Martin Gardner
Hendersonville, North Carolina

Dear Martin:

How delightful to hear from the master of mathematical puzzles himself! Mrs. Black's second hand—the one with the ace of spades—is more likely to have another ace, and here is my own explanation: There are fewer opportunities to get a particular ace than there are to get any ace at all. But each of these groups of opportunities contains an equal number of "golden" opportunities to get more aces. Therefore, the smaller group provides the greater chance of success. (Maybe you could blue-pencil this reply and send it back to me.)

He should have red-penciled it, instead. After reading it over again, I noticed that *I'd* made a mistake, so after a few weeks, I wrote a letter to "Ask Marilyn" myself and corrected the error in print:

▼▼

Dear Marilyn:

I've noticed a weakness in one of your answers. Also, I know you've been waiting for a reader to identify it so that you can correct it in print, but as it looks like that may never happen, I thought I'd better write a letter myself!

In the question, two couples are playing bridge. After a deal, an on-looker asks a player if she has an ace, and she nods; after another deal, he asks if she has the ace of spades, and she nods. After which deal is it more likely that she has at least one other ace? Or are they the same?

Marilyn vos Savant<br>New York, New York

Dear Marilyn:

That wasn't a weakness, Marilyn, that was an error! (But thanks for being so polite.) I replied that the hand with the ace of spades is more likely, and that's correct. But the explanation was wrong; it was intended to illustrate the principle behind the answer by applying it to an idealized two-ace game, but I failed to mention that.

I said there are fewer opportunities to get a certain ace than to get any ace at all. But each of these groups contains an equal number of ways to get more aces, so the smaller group provides the greater success ratio. Try this with a four-card "deck," two aces and two other cards, dealing two cards each to two players, and you'll see the principle at work.

This sort of thinking confounds people, and it isn't surprising. Although the principle isn't deep or complex, it's difficult to grasp and almost as difficult to explain, which I inadvertently illustrated myself.

On the other hand, we have an easier time understanding how probabilities work within the game of blackjack, for example, mainly because they're so pleasingly intuitive. If the chances of being dealt an ace from a full deck of cards are four out of fifty-two, it becomes clear that the chances of being dealt a second ace (in a row) is not $4/52 \times 4/52$. Since one ace (and one card) has already been dealt, the chances are $4/52 \times 3/51$. This is known as "conditional probability"—the chance of being dealt a second ace after having been dealt a first ace. And this is the reason it makes sense to memorize the cards when playing blackjack, a feat that I've never tried to accomplish. The tip of the odds is so slight that it would take longer to make money playing cards than it would take working as a croupier.

The "coincidence" of shared birthdays within relatively small groups is becoming fairly well known, but the logic and reasoning behind it aren't, which is why I decided to answer this reader's question.

▼▼

Dear Marilyn:
  I have heard that it doesn't take very many people at all in a group, maybe a dozen or so, before there are two people with the same birthday, month and day. Why wouldn't people's birthdays follow the laws of random probability?

                              Kathy Menell
                              Columbus, Indiana

Dear Kathy:
  Oh, but they do. (For the purposes of this problem, we assume that people's birthdays are spread evenly throughout the year instead of being clumped for any reason.) And according to those laws, if twenty-three people get together in a random grouping, the probability is a shade greater than fifty-fifty that at least two will share the same birthday. (Still, that

just means that half the time they will, and half the time they won't. Not such a big deal, really.)

Many people think that it would take 183 people to tip the odds past fifty-fifty, and the difficulty in "seeing" why this is wrong may come from our forgetting just how unlikely it would be for everyone to have a *different* birthday. Remember—any of them can match with any other. Envision a hundred people at a meeting. You ask all of them not only to silently choose a number between one and 365, but to try to select one that no one *else* is choosing. Wouldn't it be utterly astonishing if every single one of them did indeed pick a different number? Well, that was only a hundred people, not 183, and they were *trying*!

In the above problem, we're referring to *anyone's* birthday, not a *particular* person's. For example, my own birthday is August 11, and it would take a group of 253 people for the probability to be more than fifty-fifty that someone else would share it with me.

Although probability was mentioned in the birthday problem, it isn't really the issue. Neither is it the issue in this next problem. A psychologist at the Hebrew University of Jerusalem had been following the "Ask Marilyn" column while on sabbatical at the University of Massachusetts in Amherst and, after discussing the following question in an academic journal, she wrote to me about it:

▼▼

Dear Marilyn:

Here's a problem that may pose a challenge. Do men have more sisters than women have? (And do women have more brothers than men have?) That is, each person doesn't count himself (or herself), so, for example, in a family of two children—a boy and a girl—the boy has a sister but no brother, and the girl has a brother but no sister. Likewise, in a family of four children—two boys and two girls—each boy has two sisters but only one brother, and each girl has two brothers but only one sister.

Ruma Falk, Ph.D.
Jerusalem, Israel

Dear Ruma:

Strange as it may seem, men and women have an equal number of sisters and brothers. This is because repeatedly choosing at random one child from a family and noting how many sisters and/or brothers he or she has amounts to randomly removing one child from all families and simply counting the females and males left. Assuming an even male/female distribution of the sexes, these numbers will be about the same.

Here are the possibilities for a family of two:

| | |
|---|---|
| BB | Boy has a brother. |
| | Boy has a brother. |
| BG | Boy has a sister. |
| | Girl has a brother. |
| GB | Girl has a brother. |
| | Boy has a sister. |
| GG | Girl has a sister. |
| | Girl has a sister. |

So of the total possibilities for a family of two, two boys have a sister, and two boys have a brother; and two girls have a brother, and two girls have a sister.

After hearing that my answer had caused controversy, Dr. Falk sent the following explanation for my readers. Note that we dispense with probability in the first paragraph.

▼▼

"Let's agree to assume equal probabilities for male and female births and independence among births (whether in different families or within the same family). These are no hypothetical assumptions; they reflect biological reality fairly accurately. "Most people expect men to have more sisters than women have and more sisters than brothers. They reason that because, on the average, families have an equal number of sons and daughters, the set of siblings of men, who are themselves *not counted*, should comprise an excess of women (sisters). This apparently compelling reasoning is nevertheless wrong! Men are expected (in the long run) to have the same proportion of sisters as women have, and the same proportion of sisters as brothers.

"Asking a child from a family of, say, seven children how many brothers and how many sisters he or she has, is equivalent to picking a random family of six children and asking how many sons and how many daughters it has. In the latter case it is obvious, however, that one should expect (over many such families) the same proportion of sons and daughters. Hence, irrespective of whether the questioned individuals are males or females, they would have, *on the average*, the same number of brothers and sisters. This is the *essence* of the idea of *independence*: the probabilities of male or female childbirth in the respondent's family are *not affected* by knowledge of the respondent's gender."

So probability itself was never the issue at all. The following is another example where that's the case.

▼▼

Dear Marilyn:

A meter reader has 315 meters to read. One meter reads 476300. The next meter in succession reads exactly the same. What is the probability of this occurring?

Wendell Reed
Sedona, Arizona

Dear Wendell:

From my point of view, it looks relatively high, and here's why: Every day brings mail from people who relate similar "against-the-odds" occurrences, ranging from running into an old friend on the other side of the world to discovering a serial number matching a social security number. If you had specified the event *before* it occurred, the odds against it would be enormous. But as you only specified it afterward, it becomes simply another ordinary coincidence. That is, you only noticed whatever *did* occur, no matter what it was, and the odds of *something* occurring at some point in your life are quite high by comparison.

### Logic Loops

It can be surprisingly difficult to reason our way through seemingly easy problems. The following three are all classics.

▼▼

Dear Marilyn:

I read your column every Sunday, and some years back, I wrote to you about this question:

A man looks at a portrait on the wall and says, "Brothers and sisters I have none, but this man's father is my father's son." At whose portrait is he looking?

You said, "The man is looking at a portrait of his son," but I said and still say that he's looking at a portrait of *himself.* Will you admit that you're wrong?

Ralph Gries
Jacksonville, Florida

Dear Ralph:

Uncle! When that one first appeared, it drove people crazy, and they've never stopped writing. Let's look at it one more time.

The original answer is correct. Let's call the man speaking "John" and the man in the portrait "Mr. X." If John is an only child, and Mr. X's father is John's father's son, then Mr. X's father must be John. If Mr. X's father is John, Mr. X is John's son. That means John is looking at a portrait of his son.

▼▼

Dear Marilyn:

I've had a lot of fun with the following question: If a brick weighs three pounds plus half a brick, how much does a brick and a half weigh?

Marjorie L. Lakin
Ocala, Florida

Dear Marjorie:

Nine pounds. If a brick weighs three pounds plus half a brick, then a brick weighs six pounds (three pounds plus half of six pounds). So a brick and a half must weigh nine pounds (six pounds plus half of six pounds).

▼▼

Dear Marilyn:

You got one wrong, and I'll bet I'm not the only one to catch you on it. It was the question about the two identical tumblers. (You have two identical tumblers before you, one filled within an inch of its top with liquid A, the other filled with an identical volume of liquid B. You take a level tablespoon of A and put it into B, stirring to a perfect mixture. Then you take a level tablespoon of the mixture and put it back into the tumbler containing the A liquid. Is the final result that there is now more A in the B tumbler, or is there more B in the A tumbler? You said they're the same.)

Say tumbler A is filled with 100 percent blue dye, and one tablespoon which equals 10 percent is put into tumbler B, filled with 100 percent red dye. After mixing, tumbler B contains 110 percent of 90.1 percent red and 9.9 percent blue. A tablespoon of this mixture placed into tumbler A puts almost 10 percent of the 9.9 percent blue, or approximately .9 percent back into tumbler A, giving it a total of 90.9 percent blue, whereas tumbler B still contains only 90.1 percent red. I really enjoy your column, but I'm glad you're not mixing prescriptions.

Bill Pearson
Mayer, Arizona

Dear Bill:

I was surprised to see how many people wrote to protest my answer, but I stand by my reply.

Here's another way to illustrate the correct solution: Forget the number of spoonsful (?) and percentages and just look at the two tumblers. If tumbler A contains a certain amount of one liquid, and tumbler B contains the same amount of another liquid, no matter how many times you transfer the liquids back and forth or even whether you mix them well or not at all, as long as each tumbler ends up with an equal volume, there will be as much B liquid in tumbler A as there will be A liquid in tumbler B. This is because however much B liquid is now in tumbler A, it displaces that same amount of A liquid, which we will find (where else?) in tumbler B.


This next problem must be more counter-intuitive than I'd realized. When I published it, I thought the explanation of why the chances were so high in this case surely demolished any possible intuitive illusions, but that was not so. I continue to receive a steady stream of letters that argue that the chances are practically nil.

▼▼

Dear Marilyn:

There is a single path up a mountain. A climber starts about 6 A.M. and arrives at the top around 6 P.M. He stays there overnight, starting down the next day about 6 A.M. and arriving around 6 P.M. On each day, he travels at varying speeds—enjoying the scenery, stopping for lunch, etc. What are the chances that there was a spot on the mountain path that he passed at exactly the same time both days?

Jerry Biel
Englewood, Florida

Dear Jerry:

It's 100 percent, and here's how to visualize the proof. Imagine both the climber's trips taking place at once. The climber starts up at the same time his "twin" starts down. At some point along the way, regardless of whether one stops for lunch and the other doesn't stop at all, they will undoubtedly meet as they pass each other. That will be the place and time.

## Average Traps

This next problem is also a classic.

▼▼

Dear Marilyn:

I have asked the following "mental gymnastic" question many times, but after I give the answer, people still don't believe me: There is a race-track one mile around. If you drive around the track the first time at thirty miles per hour, how fast will you have to go around the second time to average sixty miles per hour for both times around?

Kenneth E. Wittman
Fremont, California

Dear Kenneth:

It's impossible to accomplish. To average sixty miles per hour, you'd have to drive around the track twice in two minutes, but you already used up those two minutes when you drove around the track once at thirty miles per hour.

▼▼

Dear Marilyn:

Shame on you! You erred in your answer to the racetrack question. I don't know what the new math teaches, but the old math taught me that the average of thirty and *ninety* equals sixty.

I. F. Holton
Augusta, Georgia

It should be ninety miles per hour. Race-car drivers do this sort of thing quite often. They run two or more laps to qualify. Each lap is timed. Then they are added together and divided by the number of laps.

Gerald Foye
Lemon Grove, California

What does *two minutes* have to do with anything?

John A. Tyburski
Springfield, Virginia

I'll bet you're catching lots of flak over this. Please print a correction. Who knows how many bar bets you've started?

Gil L. Dickau
Floral City, Florida

Dear Readers:

Here comes the mailman again. The original answer is correct because the length of the track was specified as the determining variable—not the length of time traveled. Envision the track straightened out, and lay the two laps end to end. It's two miles long now, right? If you then travel at thirty miles per hour for the first mile, two minutes will have elapsed by the time you reach the halfway point. If you travel at ninety miles per hour for the second mile, an additional two-thirds of a minute will have elapsed by the time you reach the end—a total of two and two-thirds minutes in all.

And when you travel two miles in two and two-thirds minutes, your average speed is forty-five miles per hour, not the sixty miles per hour we were trying to achieve! (And I'm *sure* none of my readers would ever *dream* of taking a bet in a bar, Gil.)

So now we're very well prepared for this next one, right?

▼▼

Dear Marilyn:

When you travel to work going sixty miles per hour, you arrive there early. When you travel to work going thirty miles per hour, you arrive there late. The amount of time you are early is also the amount of time you're late. How fast should you go to get to work on time?

Stan Zelinger
Mission Viejo, California

Dear Stan:

Assuming you're stubborn about leaving at the same time each day, and discounting such things as acceleration and deceleration time, you'll have to shoot out the driveway at forty miles per hour and come to a screeching halt in the parking lot at work to be right on time.

Maybe we weren't quite prepared! Let's say it's 8:00 in the morning, and your place of employment is ten miles away. You want to arrive there at 8:15. If you travel at sixty miles per hour, you'll arrive there at 8:10 (five minutes early). If you travel at thirty miles per hour, you'll arrive there at 8:20 (five minutes late). But if you travel at forty miles per hour, you'll arrive there at 8:15 (right on time).

Or let's say it's 9:00 in the morning, and your place of employment is twenty miles away. If you travel at sixty miles per hour, you'll arrive there at 9:20. If you travel at thirty miles per hour, you'll arrive there at 9:40. That must mean you wanted to arrive at 9:30, instead. And if you travel at forty miles per hour, that's when you'll arrive.

Here are two more "speed" classics. Again, the phrasing comes from the puzzle world, but they're directly applicable to the real world, if not in this precise form.

▼▼

Dear Marilyn:

Here's another puzzle for you: I lost the answer years ago, and I go nuts trying to figure it out. Smith and Jones started off together to walk to a nearby town. Smith covered the first half-mile at one mile per hour faster than Jones. She covered the second half-mile at one mile per hour slower than Jones. Jones walked to town at a constant speed all the way. Did they arrive at the same time? And if they didn't, which one got there first?

James LaBella
East Hartford, Connecticut

Dear James:

No matter what her speed is, Jones will always arrive first. This is the reason: When Smith walks one half of the distance one mile slower than Jones, she loses more than she gains when she walks the other half of the distance one mile faster than Jones. For example, let's say that Smith's "fast" speed is three miles per hour, Jones's constant speed is two miles per hour, and Smith's "slow" speed is one mile per hour. Smith will cover the first half-mile in ten minutes and the second half-mile in thirty minutes, a total of forty minutes. However, Jones will cover the whole distance in only thirty minutes. This is because when Smith is traveling fast for half a mile, Jones is traveling two-thirds as fast as she is, but when Smith is traveling slow for half a mile, *she's* only traveling *one-half* as fast as *Jones.*

▼▼

Dear Marilyn:

Suppose you have an airplane that is able to fly one hundred miles per hour (relative to the air), and you need to make a round-trip flight from City A to City B two hundred miles away. You want to make this trip in the shortest possible time. Today the winds would be a fifty miles per hour tailwind from A to B, and the same fifty miles per hour headwind from B back to A. Tomorrow the winds will be calm. Should you make the trip today, tomorrow, or would they be the same?

Mandley Johnson, Jr.
Bismarck, North Dakota

Dear Mandley:

You should go when there's no wind, and it'll take you four hours. Going with the matching tailwind and headwind will cost you an extra hour and twenty minutes.

Considering how counter-intuitive these problems are, it might come as no surprise that a pilot for a major airline wrote me to complain that the above answer was incorrect! And speaking of averaging average averages (or something like that), take a look at the following weird (but perfectly correct) conclusion.

▼▼

Dear Marilyn:

Two batters go 31 for 69 during the first month of the season, which means that both have a .449 batting average. In the following week, batter A slumps to 1 for 27. Batter B does much better, hitting 4 for 36 (the same as 1 for 9). However, both batters at that point have the same average! Batter A's 32 for 96 and batter B's 35 for 105 both equal a .333 batting average. How come?

Alvin M. Hattal
Potomac, Maryland

Dear Alvin:

Here's one way to look at it:

| | |
|---|---|
| Batter A originally has | 31 hits out of 69 "at bats." |
| He now adds these: | 1 hit out of 27 |
| For a total of this: | 32 hits out of 96 (.333 average) |
| Batter B originally has | 31 hits out of 69 "at bats," too. |
| HE NOW ADDS THESE: | 1 HIT OUT OF 27 |
| Matching A's totals: | 32 hits out of 96 (.333 average) |
| AND THESE: | 3 HITS OUT OF 9 (.333 average) |
| For a total of this: | 35 hits out of 105 (still .333) |

In other words, regardless of the order in which he actually accomplished it, batter B added a series of "at bats" to equal batter A's final .333 average *and* he added a series of "at bats" equivalent in themselves to another .333, maintaining his new average of .333.

Or, to put it another way, each batter lowered his average, from .449 to .333, but batter A, who recently did worse, lowered it faster (in only 96 "at bats") than batter B (who took 105 "at bats" to do it). But hey, who's complaining? The Mets would love either one of these guys right now.

It occurred to me that the line about the Mets might look dated by the time this book was published. But then it occurred to me that the Mets would *always* love either one of these guys!

# TWO

▼

## Lessons Set in Dollars

et's say you're a published author with one book to your credit. The nonrefundable advance against future earnings was $60,000, and although the book sold quite well, it only earned back $50,000 for you. You don't see that as much of a problem, of course. After all, even though your royalty rates were less than thrilling—it was your first book, remember—you got to keep the unearned $10,000, and the publisher seems happy, too.

So you're not surprised when your publisher offers you only $50,000 for your second book and justifies this by citing the performance of your first book, a very common practice. It makes sense, right? (And your royalty rates will improve somewhat, so you can't complain.)

But consider this. Let's suppose you'd gotten the better royalty rates the first time. What would your book have earned? $60,000? If so, then if your publishers had agreed to a *worse* deal for themselves on your last book (that is, by giving you a larger percentage of the sales, causing your first book to earn $60,000 instead of $50,000), they'd be offering you a *better* deal for your next book (that is, by giving you a $60,000 advance instead of $50,000 for your second book, in addition to the improved royalty structure).

This seems utterly wrong, and the problem is more than an issue of who stands to gain/lose or how the question is framed. It goes to the essence of logical thinking itself.

We are constantly (and illogically) being whipsawed back and forth with numbers. When interest rates are up to 15 percent, people are delighted to see the money piling up in their savings accounts, but they're miserable when they take out a loan for a new home. And when interest rates are down to 3 percent, people are delighted when they take out a loan for a new home, but they're miserable to see very little money piling up in their savings accounts.

The prescription seems obvious enough: Be happy about accumulating the cash when interest rates are high and then buy a home when interest rates are low. But for those of us without crystal balls, things seldom work out quite that conveniently. This makes us easy targets for politicians. The incumbent with *high* interest rates? "You're making money without even trying!" The challenger? "Why, you can't even buy a house!" The incumbent with *low* interest rates? "Now is the time to buy your dream home!" The challenger? "With what?"

### Economic Illiteracy

Most of us could benefit from a better understanding of economics. In 1988, according to *The New York Times*, "More than half of all high school students in a national survey could not define basic economic terms like profit and inflation. . . . only 25 percent gave the correct definition for inflation, 34 percent knew that profits equal revenue minus costs, and 45 percent could identify the term 'government budget deficit.' " A press release from the National Council on Economic Education quoted Paul Volcker, former chairman of the board of the Federal Reserve System, as saying, "While *interest* in matters of economics runs high in American minds, their *understanding* of the subject lags far behind their counterparts in other industrialized countries."

This situation appeared unchanged four years later. An overview of a survey of economic literacy conducted in 1992 by the National Center for Research in Economic Education and the Gallup Organization noted: "All survey respondents had strong opinions about economic issues despite the fact that they often had very limited economic knowledge about an economic issue."

The study assessed high school seniors, college seniors, and the general public, discovering that, for example, "the economic issue of greatest concern was unemployment. The respondents in all groups recommended a number of actions that should be taken by the federal government to reduce unemployment; yet, only about a fourth or less of each group knew the current national rate of unemployment." Also, respondents "suggested a number of actions to be taken by the federal government to reduce the federal deficit, but less than a quarter of each group knew the size of the deficit or could define a budget deficit." In addition, "only a quarter of high school students, a third of the general public, and about half of college students knew that the Federal Reserve was responsible for monetary policy. Even fewer could recognize an example of monetary policy, but a large majority of each group thought some organization other than the Federal Reserve should be responsible for conducting monetary policy."

### Choosing a Raise

When we lack knowledge, most of us operate on "gut instinct" or more enigmatically, on "intuition." But for math purposes, they're worth about the same—not much! Here's a good example:

Dear Marilyn:

Suppose you make $10,000 a year. Your boss offers you a choice. Either you can have a $1,000 raise at the end of each year, or you can have a $300 raise at the end of each six months. Which do you choose?

Ronald Gustaitis, Principal
Stepney Elementary School
Monroe, Connecticut

Dear Ronald:

Surprising, isn't it? The $300 raise continues to get better each year. At the end of one year, you'd be $300 ahead, at the end of three years, you'd be $700 ahead, and at the end of five years, you'd be $1,100 ahead. The cumulative total would be even higher ($3,500), and that's not counting interest. The $300 semi-annual raise increases your original earnings and each new salary level so often that it easily overcomes the $1,000 yearly raise, which has slower growth. (Think of it as working for six months at a time instead of a year at a time.)

This reply drove people crazy. But by the phrase "a $300 raise at the end of six months," the writer doesn't mean that your annual wage goes to $10,300. He means that in the next six months, you'll make $300 more than you did in the last six months. Here's a followup column:

▼▼

Dear Marilyn:

Are you sure about your answer to the question about the $1000 raise versus the $300 raise? You replied that the $300 raise continues to get better each year. I would love to use this example in class if I were convinced no error exists.

Arnold Barkman, Ph. D.
Department of Accounting
Texas Christian University

There must be something I am overlooking, or there must be a typographical error in the column. I would appreciate a clarification if at all possible.

Cliff Hoofman
Senator, 25th District
North Little Rock, Arkansas

One thing is clear; it pays to know exactly what the boss means in offering you a choice.

Stephen P. A. Brown
Federal Reserve Bank
Dallas, Texas

Dear Readers:

The answer is correct, and the key is that it compares two different salary periods: a one-year period with a six-month one.

Let's say it's January 1, 1995, and you choose the $1,000 raise at the end of the year. For 1995, you earn $10,000. But if you choose the $300 raise, you earn $5,000 during the first six months and $5,300 during the second six months. For 1995, you earn $10,300.

With the $1,000 raise in effect for 1996, you earn $11,000. But with the $300 raise, you earn $5,600 during the first six months and $5,900 during the second six months. For 1996, you earn $11,500.

With the $1,000 raise in effect for 1997, you earn $12,000. But with the $300 raise, you earn $6,200 during the first six months and $6,500 during the second six months. For 1997, you earn $12,700.

With the $1,000 raise in effect for 1998, you earn $13,000. But with the $300 raise, you earn $6,800 during the first six months and $7,100 during the second six months. For 1998, you earn $13,900.

With the $1,000 raise in effect for 1999, you earn $14,000. But with the $300 raise, you earn $7,400 during the first six months and $7,700 during the second six months. For 1999, you earn $15,100.


In "reply" to the above "reply," I have now accumulated many pounds of ledger sheets and letters insisting that I'm wrong. Here's another money problem, although it didn't irritate people nearly as much as the one about the raise. For this one, they simply wrote and told me politely and pleasantly that I was "all wet." ("You blew it," is usually the favorite comment.)

▼▼

Dear Marilyn:

I live in a city that has 50¢ tollbooths and 10¢ tollbooths. You can purchase a book of fifty toll tickets for $4 (8¢ apiece). When I use my tickets on the 10¢ booths, my friends say I'm losing money and that I should save them for the 50¢ booths and go ahead and pay the 10¢. I say I'm still saving 2¢. Who's right?

Danny Denton
Richmond, Virginia


Dear Danny:

You're right. With no other factors to consider, you'll do better by using a ticket at every booth, no matter what the toll. Tell your friends that your way makes all tolls 8¢, but their way makes some tolls 8¢ and others 10¢. Here are some examples.

You all go through a hundred booths, fifty at 50¢, fifty at 10¢. This costs you $8 (two books of tickets), but it costs your friends $9 (one book of tickets and fifty 10¢ tolls). Or you go through two hundred tolls, fifty at 50¢, 150 at 10¢. This costs you $16 (four books of tickets), but it costs your friends $19 (one book of tickets and 150 10¢ tolls). Or you go through two hundred tolls, 150 at 50¢, fifty at 10¢. This costs you $16 (four books of tickets), but it costs your friends $17 (three books of tickets and fifty 10¢ tolls). Your friends might be interested to note that the greater the proportion of 10¢ booths, the more *they* lose, not you!

Thomas McCormack, my editor for this book (and Chairman of St. Martin's Press), had an interesting insight into the above exchange. He said "I'll bet your dissenters picked up a signal in the question that conveyed that the motorist had only one book of tickets to dispense. The signal may have been imbedded in the phrase, 'I should save them.' I'm not saying that the readers *should* have taken the question that way, but that they *did*. In other words, the question many of your readers may have been answering was, 'If you have only one book of tickets, how do you maximize your savings?' "

This editorial comment sheds light on the thinking process. If the motorist is allowed to buy one book of tickets and *only* one book, it does indeed save him more money to use the tickets solely at the 50¢ booths and pay the toll at the 10¢ booths! But people did not write to say that I should have interpreted the question as referring to only one book of tickets. Instead, they said that my conclusion was wrong.

There's a lesson here. Perhaps these readers worked out the correct answer for a defined number of tickets—say, a book of them—and went on to assume that because the answer applied to one book, it must apply to another book, and so on, *ad infinitum*. It appears to make sense, but it's quite wrong, nonetheless. And even after reading my reply, they couldn't believe that if the motorist can buy as many books as he likes, the situation changes and he should use a ticket at *every* booth.

We can learn the necessity to *question* our assumptions, but, to do that, we first have to *recognize* our assumptions—and that's a lot harder. This is arguably the greatest logical weakness of bright people.

### Renting a Room

The next problem is a classic, and it's worth studying precisely for the reason just cited. That is, perennial favorites get that way because they're so roundly successful at confusing so many people. So let's take a closer look at this one.

▼▼

Dear Marilyn:

The following problem has been with me since childhood, so far without answer: Three people went to a hotel and rented a room for $30, each
paying $10 for his share. Later, the clerk discovered that the price of the
room was only $25. He handed the bellman five $1 bills and asked him
to return them to the three people. The bellman, not knowing how to
divide five dollars among three people, instead gave each person one dollar
and the rest to charity. Here's the question: The three people originally
paid $10 each, but each received $1 back, so they've now paid a total of
$27 for the room. Add to that the $2 that the bellman gave away, and
you have a total expenditure of $29 instead of $30. What happened to
the other dollar?

Robert T. Mann
Austin, Texas

Dear Robert:

There is no missing dollar. The total expenditure is now only $27,
accounted for by adding the $25 in the hands of the hotel clerk to the $2
in the hands of charity. In other words, the original $30 now is divided
like this: the hotel clerk has $25, the guests have $3, and charity has $2.
The error arose when an asset ($2) was added to an expense instead of
the other asset ($25), thereby mixing "apples and oranges," giving us fruit
salad instead of a correct answer.

Now try this next one, from "real life."

▼▼

Dear Marilyn:

Three of us couples are going to Lava Hot Springs this weekend. We're
staying two nights, and we've rented two rooms because each holds a
maximum of only four people. One couple will get their own room Friday,
a different couple on Saturday, and one couple will be out of luck both
nights. We'll draw straws to see which are the two lucky couples.

I told my wife we should just draw once, and the loser would be the
one unlucky couple. I figure we'll have a two out of three (66⅔ percent)
chance of winning one of the nights to ourselves. But she contends we

should draw straws twice, once on Friday and the remaining two couples again on Saturday, reasoning that a one in three (33⅓ percent) chance for Friday and a one in two chance (50 percent) for Saturday will give us better odds.

I asked her to look at it like a drawing for $10 million and asked her if she would rather have a 66⅔ percent chance of winning a single drawing or a 33⅓ percent chance in one drawing and a 50 percent chance in a second. Which way should we go?

Dave Phillips
Heber City, Utah

Dear Dave:

Actually, it's the same either way. Your chances couldn't increase relative to any other couple because you'll all be in the same drawing.

That's true, of course, and the false premise was that one way was better than another. But let's take apart the entire problem so we can get a better look at the mechanics of it.

First, envision three straws in a hat, one green marked "Friday," one green marked "Saturday" and a red marked "unlucky." Everyone draws at once, and the issue is settled. That's the kind of drawing the husband chose. Now envision the same three straws in a hat. Everyone draws at once, but the couple with "Friday" holds onto their straw for a day. Then the other two are put back into the hat, and the remaining couples choose between them. That's the kind of drawing the wife chose. But it's no different than if the second and third couples had just held on to their "Saturday" and "unlucky" straws in the first place, as above. (Except that they might switch positions with each other.)

When you draw once for a "loser," you may be trying for two nights, but you'll always split the pot. That means you have a 1/3 chance for "Friday" and a 1/3 chance for "Saturday." When you draw twice for "winners," it's the same. You have a 1/3 chance for "Friday" and "Saturday" alike, not 1/2 for "Saturday." This is because you won't be participating in the "Saturday" drawing at all if you win on "Friday," which will be one-third of the time. So you'd only have 2/3 of those 1/2 Saturday chances, and 2/3 of 1/2 equals 1/3—the same as with the other method of drawing.

## Dysfunctional Dollars

This next real-life question draws on some of the earlier "averaging" problems, and presents a very common error. In this case, a bank provided the correct answer beforehand, but in most cases, there's no one around to catch our mistakes.

▼▼

Dear Marilyn:

A bank sold me twenty checks for $1 for quite a while, which came to 5¢ a check. Later, they changed the price to ten checks for $1, making the cost 10¢ a check. But a rival bank started selling fifteen checks for $1. Now, logic tells me that because fifteen is midway between twenty and ten, the checks should cost 7½¢ a check, midway between 5¢ and 10¢. However, when I divided fifteen into $1, I find I am only paying 6⅔¢ cents a check! Why is this so? Why is the logical answer wrong?

G. W. Bartlett
Wheeling, West Virginia

Dear G. W.:

That wasn't logic; that was intuition. And it's wrong in this case because Mother Nature didn't prepare us for long division. Here's the number 100 divided by 1 through 5:

$$100 \div 1 = 100$$
$$100 \div 2 = 50$$
$$100 \div 3 = 33⅓$$
$$100 \div 4 = 25$$
$$100 \div 5 = 20$$

Look at the divisors. $100 \div 1 = 100$, and $100 \div 5 = 20$, but you know perfectly well that 100 divided by 3 (the divisor midway between 1 and 5) isn't going to equal 60 (the quotient midway between 100 and 20)! As the divisor grows larger, its incremental effect on the quotient grows smaller, and here's a way to visualize it. Let's say you're going to build yourself a new house. If you get just one more person to help, the work will be divided in half, and the benefit will be huge. But if you already have 100 people on the job, adding another will be of very small benefit.

## The Lottery

▼▼

Dear Marilyn:

Joe plays a lottery on a regular basis. The odds are 999 to one. He has played 999 times and never won. Moe plays for the first time on Joe's one-thousandth try. Are Joe's chances of winning better than Moe's?

Alan Otter
New Boston, Michigan

Dear Alan:

No, they're the same. Each lottery is a separate event, and those odds start fresh with each of Joe's (and Moe's) tries.

Blackjack is the only casino game in which the past is relevant. With roulette and with lotteries, the past has no effect on the future. Of course, this understanding is relevant in many other areas of life.

A reader wrote, "We have three young children who are all boys. Lots of people tell us to "try one more time, the odds are you'll get a girl this time!" Are the odds fifty-fifty with each pregnancy, or do you consider the sex of the children you already have?" I replied, "The odds start all over again with each pregnancy, just as they do with each flip of a coin. Relatively few people have three boys (out of three), all right, and even fewer people have four boys (out of four), but at this stage in your life as a parent, you've got a fifty-fifty chance of becoming one of the latter!"

Another reader wrote, "While attending a lecture on fire safety in the home, the speaker said, 'One in ten Americans will experience some type of destructive fire this year. Now, I know that some of you can say that you have lived in your home for twenty-five years and never had any type of fire. To that I would respond that you have been lucky.' He then went on to say, 'But that only means that you are not moving farther away from a fire, but closer to one.' Is his last statement correct? Are those people moving closer to a fire?" And I replied, "No. Moreover, people who have never had a fire are even somewhat *less* likely to have one than average, but for such nonstatistical reasons as the use of successful precautions or engaging in less high-risk behavior. For example, people who don't smoke seldom experience a mattress fire." Back to the lottery.

▼▼

Dear Marilyn:

If a lottery states that the chances of winning are one in 9,366,819, how many lottery tickets would I have to buy to be assured of winning? (This week's lottery is worth between twenty-five and forty million.)

Michael L. Laskaris
Jamaica Plain, Massachusetts

Dear Michael:

You would have to buy 9,366,819 tickets. But that's a generic answer. I don't know the various rules, regulations, and wrinkles in the many different lotteries. For example, I wouldn't want to mislead you into thinking that if you could simply purchase all the above tickets for $9,366,819, you would be assured of a fabulous profit by winning a pot of more than $25 million!

Not surprisingly, people have tried to do just that, and that's the obvious weakness in holding a lottery in which the prize is worth more than the price of all the tickets. (Which is why it isn't done, of course. Or at least *shouldn't* be.) In February of 1992, an Australian investment group attempted to buy all possible combinations of six numbers from one to forty-four—a total of 7,059,052 tickets—for a Virginia state lottery that had a $27 million jackpot.

This particular lottery was a good choice. It had the biggest grand prize in the country that weekend, and the amount of money needed to cover all the bets was less than a third of the amount needed to "buy" the lottery in a state like New York, which has fifty-four numbers instead of forty-four. Moreover, the second-, third-, and fourth-place prizes made the grand prize worth even more. The only danger—but it was significant—was the possibility of having to split the jackpot with other winners. (Some numbers are very popular, and there may be hundreds of ticket-holders for them.)

Just the logistics of accomplishing such a purchase (and filling out all those play slips) are mind-boggling. Imagine the line that would accumulate behind anyone buying seven million tickets! (In fact, the Australian group spread out its purchases and used other aids to buying so many tickets, but plenty of lines did accumulate, and lottery officials received numerous complaints.)

At $1 a ticket, the group managed to buy about five million tickets before running out of time. The situation must surely have weighed heavily on their nerves. After spending $5 million, it had "only" a 5/7 chance of winning. Before a winner came forward, lottery officials already were considering how to block future reruns of this unwelcome show. Worse, at least for the hopes of the investment group, the officials were even debating ways to invalidate that particular lottery.

Regardless, I only wish I could have been there to see some sweaty fellow, his jacket on the back of his chair and his tie loosened, searching through little pieces of paper until he found "The One." And indeed, the Australian group did find it. Even luckier for the investor, there was no one else to share the prize. (Not that this win was going to make anyone rich. Without including commissions and expenses, if a member of the group bet $1,000, he or she would only get back about $4,000—a fabulous rate of return on the "investment," but without complete control of the situation, it was still wildly speculative.)

The payout was postponed while irked lottery officials considered their options, including outright defaulting on the rules they themselves had made. They finally decided to award the prize to the Australian syndicate, which consisted of some twenty-five hundred investors, most of them Australian, but some from the United States and Europe, too. Kenneth Thorson, the state lottery director, said that despite the fact that the syndicate had broken the rule requiring the completion of each transaction at the site of each ticket purchase, thus casting doubt on the validity of the winning ticket, "We believe that all doubts must be resolved in favor of the player." (Not surprisingly, he also mentioned the likelihood of a lawsuit and the state's weak case.)

We all know though, that when we buy just one ticket, the lottery is nearly impossible to win—but just *how* impossible is it? After all, we see plenty of newspaper reports about winners, don't we? In a "Mathematical Recreations" column that appeared in *Scientific American* (March 1990), A.K. Dewdney outlined an amusing example of his own. "What are my chances of winning a lottery?" he asked. "Consider the case of the six-number ticket. Players buy a ticket on which they select six numbers from one through a hundred." According to Dewdney's calculations, ". . . the total number of lottery outcomes . . . barely fits in my calculator:

1,192,052,400. If I buy one ticket . . . the chance of winning the lottery is therefore about one in one billion, and the probability is approximately .000000001!" He went on, "My chance of winning the lottery is about two times smaller than the chance that the next meteorite to hit the earth will land within the square kilometer that has me at its center."

Dewdney commented that his analogy was chosen to surprise people—to combat "number numbness," as he put it—and he justified it as follows: "The surface area of the earth is approximately 510,000,000 square kilometers. The chance of the meteorite falling within my particular square kilometer is therefore one in 510,000,000: the probability is about .000000002. Because this probability is about two times greater than the chance of winning the lottery with one ticket, I would have to purchase two tickets to even the odds."

His calculations appear to make sense at first reading, but how can this actually be the case? We see stories about lottery winners "all the time," and everyone even seems to have an Aunt Gertrude who once won a lottery. But if it's twice as easy to (almost) get hit by a meteorite, then where are all the meteorite stories?! (And if no one actually gets beaned, don't meteorites ever land at busy intersections or barely miss a cruise ship?)

There's an explanation, of course. Even if we don't count the many Aunt Gertrudes who win $5 and instead count only the ones who are grand-prize winners, the fact remains that numerous lotteries are held. But how many meteors, those chunks of extraterrestrial matter that we call "shooting stars" when they enter the Earth's atmosphere, survive the incandescent, disintegrative process to reach the surface as meteorites? If we also don't count the micrometeorites that I like to call "stardust," there would have to be as many "falling stars" that you can later put into your pocket (albeit a big one) as there are lotteries for a similar number of headlines to be generated.

But there aren't. So as far as the validity of the meteorite example is concerned, we may as well say that your chance of winning the lottery is about two times smaller than the chance that the next alien baseball to hit the earth will land within the square kilometer that has you at its center. Well, that's true, I suppose—*if* an alien ever hits a homer of that magnitude our way!

(Dewdney later wrote to explain to us that he was indeed including micrometeorites, so don't write to him to complain! His mailbox is already full.) Of course, we can also take a closer look at the math—lottery chances are seldom as bad as one in a billion—but math isn't the point of this book. Mathematical intuition is, though. What seems simple—and obvious—at first, may be complex. Or it can be just the opposite of what we'd expect. This next problem is deeper than it sounds at first, so be forewarned. (There's going to have a quiz afterward!)

▼▼

Dear Marilyn:

If I only have $10 to spend on the next ten drawings of the lottery, and the odds against winning are thirteen million to one (each time), wouldn't it be most advantageous to bet it all at once? It seems that lowering the odds for one drawing would give a much better chance of winning than betting $1 on ten different lotteries. Correct? Incorrect? This had been driving me nuts!

Angeline O'Brien
Racine, Wisconsin

Dear Angeline:

It's the same either way. Although it may seem at first that buying the ten tickets all at once gives you a better chance because the betting pool will be smaller (thirteen million in just that one lottery instead of ten times that number in all of them), don't forget that over the course of the ten lotteries, there will also be ten prizes, not just one.

We also need to consider here the meaning of the word "advantageous." Using an extreme case as an example, let's say there are ten drawings made up of ten tickets each. The prize in each drawing is $10, and tickets may be purchased for $1. If you buy one ticket for each of the ten drawings, you've spent $10, but you might lose each time. On the other hand, if you buy ten tickets for only one drawing, you've spent an equal amount, but you'll win for sure. What's the meaning of this?

Well, it's close to the meaning of the term "hollow victory"! Did "advantageous" mean increasing your chances of winning? Or increasing your chances of winning more than you lose? Spending $10 to win $10 means that you'll be sure to win nothing at all.

(I was just kidding about the quiz.)

▼▼

Dear Marilyn:

A lottery ticket is a winner when its six numbers match those drawn from a forty-number pool. A ticket purchaser, basing his selection on family members' birth dates, thus restricted his possible winning combinations to number one through thirty-one. Did he decrease his chances of winning by his eliminating numbers thirty-two through forty from his possible choices? Your answer could quite possibly prevent World War III from breaking out in our family.

Lewis B. Hall
Baltimore, Maryland

Dear Lewis:

I hope I'm not too late. No, there was no change whatsoever in the chances of winning. No matter which numbers were selected, all the others were eliminated, and the numbers *he* eliminated were no more likely to be chosen than the numbers anyone *else* eliminated. I hope that settles it. And if World War III *does* break out, we'll all going to hold you guys responsible.

There's yet another angle to be considered—although not like the "hollow victory" one. I should have mentioned this factor in my previous reply.

▼▼

Dear Marilyn:

You said once that a ticket-purchaser who based his selection on the birthdates of family members didn't decrease his chances of winning and that the numbers he eliminated were no different from the numbers anyone else eliminated. However, he eliminated these numbers on a systematic basis, and no one else did. This must make some difference. If not, could you please show me the error in my thinking?

Michele A. Leimgruber
Demarest, New Jersey

Dear Michele:

It can make a difference, but not in his chances of winning, and the kind of difference depends on how the lottery is constructed. If the prize is $1 million, a million different tickets are sold, and the winner is to be drawn from among those, it doesn't matter which ticket he purchases.

But let's say those million people get to *choose* their ticket numbers, the prize to be split among any duplicates. Then 999,999 people pick number 7, and you pick number 11. The winner is drawn from the numbers one through one million. While the chances are equal that number 7 and number 11 will win, you'd have a much better payoff with number 11 than all those people sharing number 7. In this kind of setup, avoiding "popular" numbers is a wiser bet.

The answer to this next question is far more surprising.

▼▼

Dear Marilyn:

My son-in-law and I have a disagreement on the lottery. He thinks the chances of winning the lottery are just as good when you pick the numbers in sequence (1-2-3-4-5-6) as they are scattered (3-6-15-20-39-43). But I've played the lottery for many years, and I've never heard or seen six numbers drawn in sequence. Please settle our dispute.
                              Richard J. Ciesielski
                              Fort Wayne, Indiana

Dear Richard:

The chances are the same. The reason people remember fewer winning "sequence" numbers is that, depending on the lottery rules, there are usually fewer of them than there are "scatter" numbers. However, when you buy a lottery ticket with a scatter number, you don't get *all* the scatter numbers; you only get *one*. This means that while the chances that *some* scatter number will be chosen are greater than some sequence number, the purchaser of a *particular* scatter ticket doesn't get those same chances.

Here's a way to illustrate it. Let's say there are six numbers in the lottery and two numbers on a ticket, all sold. The tickets would look like this, sequences bracketed.

| [1–2] | [2–1] | 3–1 | 4–1 | 5–1 | 6–1 |
| 1–3 | [2–3] | [3–2] | 4–2 | 5–2 | 6–2 |
| 1–4 | 2–4 | [3–4] | [4–3] | 5–3 | 6–3 |
| 1–5 | 2–5 | 3–5 | [4–5] | [5–4] | 6–4 |
| 1–6 | 2–6 | 3–6 | 4–6 | [5–6] | [6–5] |

There are twice as many scatter numbers (twenty) as there are sequence numbers (ten), but picture them all (thirty) written on slips of paper and dropped into a box. The chances of picking out any *one* ticket are still one in thirty.

But what about this next question?

▼▼

Dear Marilyn:

I play the same five numbers without fail in the lottery. I know that the mathematical odds of winning start over with each drawing, but do you agree that sooner or later, nearly every combination of numbers will have come up? (Hope so!)

Peggy Hafner
St. Paul, Minnesota

Dear Peggy:

I see what you're getting at. Yes, given an unlimited amount of time, every number eventually will be chosen. But the flaw in this reasoning is that the lottery doesn't remember which numbers have already won once (or twice or three times) and thus know to avoid them, so no particular number has a better chance the next time. In fact, other numbers are more likely to win two or even many times before every single number wins even once.

These next two problems are favorites of mine, but the logic behind the first is far easier to see. (Read carefully—there'll be a quiz afterward.)

▼▼

Dear Marilyn:

Here's one that drives me nuts: If we're playing roulette on a wheel that has thirty-eight slots, the probability of the ball dropping into any one specified slot is one out of thirty-eight, right? But let's say we use two balls at once. What's the probability of both balls dropping into any slot together?

Mike Guider
Richmond, Virginia

Dear Mike:

It's still one out of thirty-eight. The first ball "specifies" the slot, and the second ball has a one out of thirty-eight chance of dropping into it.

Now, would you like to hear a dandy "sucker" bet? This one tops them all, as far as I'm concerned. It comes from Martin Gardner.

▼▼

Dear Marilyn:

Let's say we're playing roulette, and I offer you a bet. You can pick any triplet of black and red—say red/red/black or red/black/red. Then I'll pick a different one. At the starting point, we'll watch each spin of the roulette wheel until one of our triplets appears as a run. If yours comes first, you win. If mine comes first, I win. Even chances, right? But I'll give you three to two odds! When you win, I pay you $3, but when I win, you only pay me $2. We'll play as many times as you like, and you can always have the first choice. Will you take the bet?

Martin Gardner
Hendersonville, North Carolina

Dear Martin:

What a great "sucker" bet that would be! No, your chances of winning would range from 2/3 to 7/8 depending on what triplet I choose. You would always be able to choose a triplet with a better chance of winning. Let's use the 7/8 chances as an example because it's the most obvious. There are eight different triplet combinations, and let's say I choose black/black/black. If it appears at the start, I win, and that'll happen one-eighth of the time. But before it appears any time afterward, it would have to be preceded by a red. So if you choose red/black/black every time I choose black/black/black, you'll win seven-eighths of the time!

### An Impossible Average Classroom

Okay, here's the quiz. (What? Your intuition told you I was kidding? But don't you remember what I told you about your intuition?!) Actually, it's a letter from me to you.

▼▼

Dear Readers:

Let's suppose you've chosen the wrong raise and eventually spent your last dollar on yet another losing lottery ticket, so you've decided to go back to school and learn a useful (if not lucrative) trade like writing books about counter-intuitive thinking that confound people who even try to read them, let alone those who stubbornly insist on understanding them, too.

After looking through all the catalogs, you choose a school that advertises an average class size of only twenty students for their 100 courses, ranging in size from an intimate five students to a lecture hall with two hundred. Most of the courses, they add, have ten students or fewer.

However, after arriving on campus and talking to people, you discover that most students find themselves in larger classes most of the time. As a matter of fact, the average student's class size is more than fifty-three!

Should you demand your money back?

Marilyn vos Savant
New York, New York

No. Let's take an extreme example and say a school offered only three courses—reading, writing, and arithmetic. Ninety-nine students registered for reading, two signed up for writing, and one took arithmetic. The school then advertised an average class size of thirty-four. (99 + 2 + 1 = 102 ÷ 3 = 34) But only the arithmetic student could have had an average class size of thirty-four, and only if she took all three classes. (If she took only arithmetic, she had an average class size of one; if she took arithmetic and writing, she had an average class size of 1.5.) A writing student who took only writing would have an average class size of two; if he took writing and reading, he had an average class size of 50.5. Thus, nearly all of the students will have an average class size of ninety-nine. Why? Because there are more students in the larger classes!

Let's look back at our example. We said that the school advertised an average class size of only twenty students for their one hundred courses and that the classes ranged in size from five students to a lecture hall with two hundred. Most of the courses, we added, had ten students or fewer. Here's one way that could be the case:

| | | |
|---|---|---|
| 10 courses with | 5 students = | 50 students |
| 10 courses with | 6 students = | 60 students |
| 10 courses with | 7 students = | 70 students |
| 10 courses with | 8 students = | 80 students |
| 10 courses with | 9 students = | 90 students |
| 10 courses with | 10 students = | 100 students |
| 37 courses with | 30 students = | 1,110 students |
| 1 course with | 100 students = | 100 students |
| 1 course with | 140 students = | 140 students |
| 1 course with | 200 students = | 200 students |
| 100 courses with | | a total of 2,000 students |

With the above course listing as an example, the class size is indeed twenty students ($50 + 60 + 70 + 80 + 90 + 100 + 1,110 + 100 + 140 + 200 = 2,000 \div 100 = 20$). The classes also range in size from five to two hundred, and most of them (60 percent) have ten students or fewer.

Let's make it quicker to explain by also assuming that every student takes only one class. (If you'd like extra credit here, you can figure out why this algorithm works for students who take more than one course, using our three-course school as an example. The average student always has an average class size larger than the average class. The only time the two "average classes" are equal is when all the classes are the same size. This time, literally—go figure.)

So you arrive on campus and find that the average student's class size is more than fifty-three! How did we determine that? Using the above course listing, here are the numbers. (If there were ten courses with five students each, then five students were with five students ten times, or 5 × 5 × 10 = 250.)

| | | | |
|---|---|---|---|
| 50 students find a class size of | 5 = | 250 total |
| 60 students find a class size of | 6 = | 360 total |
| 70 students find a class size of | 7 = | 490 total |
| 80 students find a class size of | 8 = | 640 total |
| 90 students find a class size of | 9 = | 810 total |
| 100 students find a class size of | 10 = | 1,000 total |
| 1,110 students find a class size of | 30 = | 33,300 total |
| 100 students find a class size of | 100 = | 10,000 total |
| 140 students find a class size of | 140 = | 19,600 total |
| 200 students find a class size of | 200 = | 40,000 total |
| 2,000 students find a total class of | | 106,450 |

And because 106,450 ÷ 2,000 = 53.225, the average student will find more than fifty-three in a class (including him/herself).

When you reflect on all this, keep in mind that this isn't a number trick or some kind of illusion. It is a real-life situation and underlies the reason that experiences will so often seem at odds with projections. Such discordance occurs in all areas of life, most devastatingly in those with financial implications. Further, understanding how this works is a basis for understanding the reason that statistics can be so utterly misleading—regardless of whether they're actually intended to deceive or not. And that's the subject of the next part of this book.

### A True Paradox

If you're inclined to believe that, as a result of reading this book so far, you're never going to be foiled by another poll or fooled by another politician, think again! Before we continue, I want you to have a look at a real paradox—a crusher of confidence, a destroyer of dreams, and a master of men. In a slightly more quotidian vein, John Allen Paulos states, "Variations of it account for some of the 'grass is always greener' mentality that frequently accompanies the release of statistics on income."

▼▼

Dear Marilyn:

I am confronted with selecting an envelope from two envelopes, knowing one envelope (but not which one) contains twice as much money as the other one. I find $100 in my first selected envelope. Should I switch to the other one to improve my worldly gains?

Barney Bissinger
Hershey, Pennsylvania


Dear Barney:

This is a dandy paradox. While it appears as though you should switch because you have an even chance for $200 versus $50—which any gambler would grab—it actually makes no difference at all. Those even chances would apply only if you could choose one of *two* more envelopes, one with $200 and the other with $50. As it is, there's just one more envelope sitting there, with either twice the amount you've already seen or half of it. And you knew that would be the case before you even started. So when you opened the first envelope, you didn't gain any information to improve your chances.

This can be illustrated by noting that the logic that causes you to switch (because you appear to have an even chance for $200 versus $50) will lead you to switch *every* time (no matter what you find in the first envelope), making the second envelope just as randomly chosen as the first one!


Odder still, consider the same problem phrased so that one envelope contains a *million* times as much money as the other one. You find $1 in your first selected envelope. (That is, the other envelope holds either a $1 million or a bit of copper dust.) Should you switch?!

I'll bet that last question will bring on a sleepless night. Then again, maybe I was just plain wrong to include my address in this book.

# Part Two

HOW NUMBERS AND
STATISTICS CAN MISLEAD

# Three

▼

## Misunderstanding Statistics

Is a person with an I.Q. of 200 "twice as smart" as a person with an I.Q. of 100? No. Even if you give intelligence tests more credit than they deserve and insist that human intelligence can be adequately quantified in a truly meaningful way, an I.Q. of 200 *still* wouldn't be "twice" an I.Q. of 100. But why? Our intuition tells us that two hundred of something is twice as much as one hundred of something. And that's one of the reasons that statistics are so devoid of meaning so much of the time. Even the most obvious "conclusions" are often deeply flawed. This is one of the reasons that our information-oriented society is in so much turmoil about how to handle all the problems that are described by numbers in one way or another. And that's nearly all of them.

- the national debt
- welfare reform
- crime
- education
- inflation
- health care
- unemployment
- poverty
- pollution
- income taxes

And there are many more, of course. Nearly every time we read an article about one of these problems, statistics will be used to bolster the argument—whatever it is. And the majority of the time, those statistics will be misleading, at the least. But let's leave most of those problems for discussion in the last part of this book and go back to the subject of intelligence testing for now. Here's an example of how easy it is to mis-

63

interpret a simple statistic. Is a person with an I.Q. of 200 "twice as smart" as a person with one of 100?

### Normal Distributions

Many human characteristics follow what we call a "normal" distribution (known as "Gaussian" to mathematicians), and I.Q. is one of them. It means that most people measure somewhere in the middle, and the number of people at any given point decreases from that midrange as we approach either extreme, both high and low. That is, fewer and fewer people are very bright or very dull, and hardly anyone is at the top or bottom. This lineup will graph in a "bell curve," which is shaped roughly like a bell and symmetrical around the middle point. In the case of I.Q., that middle point is a score of 100—the most common I.Q.

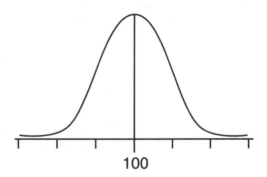

100

Now, what does that score of "100" mean? A hundred *what*? Inches? Pounds? No, as we soon realize, the "100" is an arbitrary figure—the assignment of a round number to represent the intelligence of an average person. Then we can stretch our bell curve around it in any way that suits our convenience, depending on how many questions are on our test and how many points we decide to award for each correct answer.

On most tests, a score of 50 represents the very dull and 150 represents the very bright, and designers of intelligence tests try to adhere to those standards for the same reason that dress designers do—to avoid comparison chaos. But it doesn't really matter. When the extremes are a little closer together, the bell curve thickens somewhat in the middle; when they're father apart, the opposite occurs. Either way, a person's percentile ranking within the population will remain the same.

But is an I.Q. of 200 "twice as smart" as one of 100! Well, what if an early test designer graphed his results from 90 to 110 instead, and *that* had become the standard? People's I.Q. scores would be routinely read in tenths (like body temperature), and the topmost scorers would hover around, say, 110.* So even though the same people are involved, no one would think a person with an I.Q. of 110 is "twice as smart" as one with 100!

Let's also note here that we're not referring to tests with "ceilings." Scores on such tests do not graph as a bell. Instead, the high end is cut off because the test is not hard enough to make the numbers of higher and higher scores grow increasingly rare. That is, too many people will bump into the "ceiling." Comparing a score of 150—on a test with a ceiling of 150—to a score of 150—on a test with a ceiling of 200—is also an error, but of still another kind.

But not all distribution curves are normal. Adding to the general confusion, obfuscation, and obduration, is the fact that statistics are often unaccountably bumpy, intermittent, and lopsided.

Here's where the statistical concepts of "mean," "median," and "mode" begin to make their presence felt. All three are deceptively called "average," sometimes intentionally and sometimes not. However, they describe very different phenomena. If we don't know which "average" is being used, we're at the mercy of our source of information, whether he or she is a well-intentioned television newscaster or a seasoned politician.

### Mean, Median, and Mode

The "mean" is the arithmetical average—the one with which we're most familiar. If, for example, we're discussing the income of a population, we total the income of all the people, then divide that figure by the number of people, and that's the "mean" income. Maybe. Including children can be misleading, so we might want to total the income of all the people, then divide by the number of adults, and *that's* the mean income. Well, maybe. Including unemployed people can be misleading, so we might

---

*In fact, the actual number would be well defined by what we call "standard deviation," a measure of the relative spread of a statistical sample taken from a normal population. Comparing a score of 150—on a test with a standard deviation of 15—to a score of 150—on a test with a standard deviation of 30—is also an error, but of a different kind.

want to total the income, then divide by the number of adults who are employed, and *that's* the mean income. But wait. Shouldn't the unemployed people be counted as having zero income? If so, what about the voluntarily unemployed?

But let's suppose there are no decisions like those to make. Suppose we just want to know the average income of the employees of the Brand X Corporation. Here's the accounting of salaries:

| | |
|---|---|
| $470,000 | earned by the president of the company |
| 100,000 | earned by his wife |
| 80,000 | earned by each of his wife's three brothers |
| 50,000 | earned by his wife's best friend from high school |
| 30,000 | earned by his plant manager |
| 25,000 | earned by each of the six production workers |

They must whistle while they work! The average employee of Brand X Corporation earns $80,000! (The total annual payroll of $1,040,000 divided by thirteen employees equals $80,000.) The president's wife can truthfully say that her brothers earn no more than the average employee. And think of how happy everyone will be if the president earns $990,000 next year. Why, the earnings of the average employee will jump to $120,000! (The total annual payroll of $1,560,000 divided by thirteen employees will equal $120,000.) And the president's wife could now truthfully say that she earns less than the average employee.

Not that the use of the mean is all bad; quite the contrary. We use it daily. Let's suppose that the thirteen employees of Brand X Corporation consumed 182 aspirin tablets last week. Then let's suppose that we're going to buy the supply for next week. Of course, a calculation is unnecessary—we'll just purchase the same amount. But the underlying reasoning goes like this:

Those 182 tablets divided by thirteen employees equals a mean of fourteen aspirins per employee. Assuming that we anticipate the same number of employees to be with the company the following week, we get a projected aspirin consumption of 182 tablets. Obviously, this makes sense. That's what the employees consumed in the previous week. We'll get back to this example, but in the meantime, read on.

The "median" is another kind of average that is commonly used by those who want to influence our opinion. Strictly speaking, it is the numerical middle. In the example of the Brand X Corporation, the median income is $30,000. That is, six employees earn more than that, and six employees earn less. Most people believe that the arithmetical average should be used to calculate all averages and are likely to reject the possibility that the numerical middle is ever representational, but it's clear that this notion is flawed. After all, the mean income of the Brand X employees is $80,000, and the median income is $30,000. In this case, then, the median probably comes closer to telling us what we want to know.

Not that the median is all good. On the contrary, let's get back to that example of how the thirteen employees of Brand X Corporation consumed 182 aspirin tablets last week. Here's the breakdown, ranked in order of aspirin tablet consumption:

| | |
|---|---|
| 40 | taken by the wife's brother Tom |
| 40 | taken by the wife's brother Dick |
| 40 | taken by the wife's brother Harry |
| 30 | taken by the wife |
| 20 | taken by the president of the company |
| 5 | taken by his plant manager, Doc |
| 2 | taken by the production worker Grumpy |
| 1 | taken by the production worker Sneezy |
| 1 | taken by the production worker Bashful |
| 1 | taken by the production worker Happy |
| 1 | taken by the production worker Sleepy |
| 1 | taken by the production worker Dopey |
| 0 | taken by the wife's best friend from high school |

The median consumption of aspirin tablets is two. That is, six employees take more than two, and six employees take fewer than two. But if we base our purchase of next week's supply on the median consumption of two aspirins times thirteen employees, we'll buy only twenty-six tablets, clearly too few.

The "mode" is yet another kind of average that is a convenient tool of manipulation. Simply put, it is the most common of the numbers cited. In the Brand X Corporation, the modal number of aspirins taken is one.

That is, more employees (five) take one aspirin than take any other number of aspirin. The next most common number of aspirin taken is forty (three employees). So if the five employees who take one aspirin take a brand called Exasperin, the manufacturer can claim that "more people take Exasperin, than any other brand," at least at the Brand X Corporation. Suppose, then, that similar numbers are obtained in a much broader study. Five *thousand* employees take one aspirin, and three *thousand* employees take forty, and so on. Again, the manufacturer can claim that "more people take Exasperin than any other brand," and in a comprehensive survey, yet. But what if *only* those employees took Exasperin, and we're considering buying stock in the Exasperin Corporation? Wouldn't it be interesting to discover that the sales of Exasperin accounted for less than three percent of the market?!

Not that the mode is all bad. On the contrary, let's go back to our aspirin example. In the Brand X Corporation, the modal number of aspirins taken is one. But the modal *aspirin* taken is something else entirely. Let's suppose that we're going to purchase the next week's supply of aspirin for the Brand X Corporation, and we've wisely decided to buy at least 182 tablets (and perhaps a few more in the event that the annual report is nearing publication). If we limit ourselves to only one product, which should we select? Let's say we've already decided not to consider the cost of the product, concerning ourselves with employee satisfaction, instead. Here's how their favorites from last week line up:

| | |
|---|---|
| 40 | Roboprin taken by the wife's brother Tom |
| 40 | Roboprin taken by the wife's brother Dick |
| 40 | Roboprin taken by the wife's brother Harry |
| 30 | Nouveauprin taken by the wife |
| 20 | Exasperin taken by the president of the company |
| 5 | Exasperin taken by his plant manager, Doc |
| 2 | Exasperin taken by the production worker Grumpy |
| 1 | Exasperin taken by the production worker Sneezy |
| 1 | Exasperin taken by the production worker Bashful |
| 1 | Exasperin taken by the production worker Happy |
| 1 | Exasperin taken by the production worker Sleepy |
| 1 | Exasperin taken by the production worker Dopey |
| 0 | taken by the wife's best friend from high school |

There are two modal choices, and either one may suit our purpose. Roboprin is the aspirin consumed in the greatest number (forty each by the wife's three brothers, a total of 120 tablets), but Exasperin is the aspirin consumed *by* the greatest number (the president of the company, his plant manager, and the six production workers—a total of eight employees). That is, we can select either Roboprin (because it will be taken in the greatest number) or Exasperin (because it will be taken *by* the greatest number). Or, if we come to our senses in time, we can always select Nouveauprin (perhaps a wise choice, regardless).

Misinterpretation and disinterpretation of averages are worse than common—they're both rampant and ruinous to our understanding. Not long ago, I read an article about how women who have been raped took an average of nearly a year to bring themselves to tell another person about it. The psychologist who wrote the article regarded this as overwhelming evidence of their fear of a court system that would humiliate them and a fatalistic attitude about women's continuing victimization.

I was skeptical enough by this point, but when she went on to relate the one-year time lag to seasonal changes (such as the smell of spring or the leaves turning color) that brought back vivid memories of the crime, I put the magazine aside and tracked down the source she cited. Sure enough, she'd taken the reported *averages* and hadn't looked at the data that produced them. It turned out that the great majority of women in the study had reported being raped to the authorities almost immediately, but a few were so ashamed about it that they didn't admit it to another soul for ten or twenty years. Add these two groups together, and what was the "average" reporting time? One year. (An analogous example would be if nine women reported a rape to the police immediately, and one woman waited ten years, then told her best friend. Ten years divided by ten women equals an average reporting time of one year.) But not a single woman *actually* waited a year! Far from it, in fact. The situation was turned into fiction, then crowned with a groundless conclusion.

With all these caveats, it shouldn't surprise us that *none* of the three types of average is necessarily germane to our interests. We can easily drown crossing a body of water that averages two feet deep, whether that average refers to the mean, the median, or the mode. But what more commonly escapes our attention is that neither do *other* results of statistical surveys always—or even typically—have relevant significance, regardless of how carefully the data is gathered and reported. The point, then, is not that statistics "lie"—which they do—but just how seductive a presence are they. Consider:

- Even when irrelevant, statistics get our attention, and the fewer of them that are cited in a situation, the more pronounced is the effect. A single statistic is like a television set that's glowing over the bar—almost impossible to ignore.
- And when statistics *are* relevant, we often limit ourselves to using only the ones that are available, as though the ways in which a situation happens to be enumerated define the parameters to be considered.
- But relevant or not, it is even more difficult to ignore statistics selectively than to ignore them altogether. That is, given half a dozen statistics, we are likely to use them all in our decision.

We're even likely to give them all roughly equal weight, although this latter error tends to decrease as the sheer volume of statistics increases, perhaps not surprisingly. (So many statistics—so little time!)

### Numb Numbers*

- Average level of testosterone in the saliva of male trial lawyers, in nanograms per deciliter: 6.7
- Average level in the saliva of male nontrial lawyers, in nanograms per deciliter: 5.7
- Average number of shopping carts stolen from American supermarkets each hour: 38
- Estimated average number of seconds it takes a New York City thief to break into a locked car: 27
- Percentage of Americans who think "espresso" is an "overnight delivery system": 7
- Percentage of Americans who don't know how long it takes to hard-boil a chicken egg: 74
- Percentage of Americans who say they are "excellent" or "very good" cooks: 63

*The preceding statistics were taken from "Harper's Index," a registered trademark of *Harper's* Magazine. Original sources appear at the end of the book.

- Maximum speed of particles expelled during a sneeze, in miles per hour: 103.6
- Percentage of Canadians who expect their country to become a part of the United States in the next fifty years: 37
- Percentage of Americans who believe that the president can suspend the Bill of Rights during wartime: 23
- Percentage of male college students who believe that life is "a meaningless existential hell': 27
- Change, since the Berlin Wall fell in 1989, in the percentage of western Germans who say they suffer indigestion: +5
- Change, since then, in the percentage of eastern Germans who say they suffer indigestion: −12
- Percentage of TV viewers who say they wouldn't give up watching TV in exchange for any amount of money: 25
- Percentage of Super Bowl viewers who do not live in the United States: 88
- Average number of days each year that an American is in a bad mood: 110
- Average percentage increase in the bounce of a golf ball that has been passed through an electron beam accelerator: 5
- Average speed of Heinz ketchup, from the mouth of an upended bottle, in miles per year: 25
- Chances that a funeral procession in Taiwan includes a stripper: 1 in 3
- Average amount an American would be willing to pay to see live dinosaurs: $15
- Percentage of Americans who say they would not be willing to pay anything at all: 55
- Ratio of the number of wild turkeys in the United States to the number of wild-turkey hunters: 2 to 1

Those are "fun" statistics, meant to provoke a smile—and they do. But before we move on to more serious numbers, let's learn a lesson from these more playful figures.

For instance, how in the world does anyone come up with the average number of shopping carts stolen each hour? Simple. You start with the lackluster facts, according to the Food Marketing Institute in Washington, D.C., that there are about 30,400 supermarkets nationwide and that the average one loses eleven carts per year. Multiply those numbers, and you find that 334,400 shopping carts were stolen last year. Then you divide that by 8,760 (hours in a year) to get the "average number of shopping carts stolen from American supermarkets each hour: thirty-eight." It's a much snappier presentation of otherwise dull figures. And it's honest, too.

Similarly innocent is the "percentage of Super Bowl viewers who do not live in the United States: 88." It's both surprising and amusing to envision the whole world glued to American television, but when we stop to consider that the Super Bowl is broadcast to much of the rest of the planet, and that the U.S. population comprises only a small portion of the Earth's population, the number grows far less surprising. Even if the popularity of the Super Bowl is *huge* in the United States (and it is), if our country is dwarfed by all the other countries put together (and it is), the rest of the world can be largely uninterested in the Super Bowl (and it is), and our numbers will be swamped by their numbers.

According to A.C. Nielsen Media Research, about 91 million Americans watched the Super Bowl in 1993, and according to the National Football League's "guesstimate," maybe 750 million Earthlings watched it, all told. That means about 12 percent of the total estimated viewers were Americans—ergo about 88 percent weren't, which dovetails with the original statistic. But a different way to describe those same numbers is that with 91 million American viewers out of a total United States population of 250 million, about 36.4 percent of the American population tuned into the 1993 Super Bowl. And with 659 million non-American viewers (750 million viewers in the world minus the 91 million viewers in the U.S. = 659 million) out of a total non-United States population of 5,134 million (5,384 million people in the world minus the 250 million people in the United States = 5,134 million), about 12.8 percent of the non-American population tuned in to The Super Bowl.

In short, according to the above sources, 36.4 percent of the Americans watched the Super Bowl, but only 12.8 percent of the non-Americans did. To make the opposite point of the original statistic, then, we can simply divide 36.4 by 12.8 and arrive at the conclusion that the Super Bowl is about three times as popular in the United States than in the rest of the world. So they're not glued to American television, after all (yet). But the foreign viewership still seems like a high number, doesn't it? After all, 12.8 percent equals about one out of every eight people. How was the worldwide viewership statistic "guesstimated"?

First, we telephoned Nielsen, but they have no means of gathering information outside of this country, either directly or indirectly, and they didn't have firsthand knowledge of any American or foreign organizations that do. Next, we telephoned ESPN International, which carried much of the foreign coverage, and asked if there was any way at all to guess how many non-Americans watched the Super Bowl. Not to their knowledge, they said. But there were surely plenty of *possible* foreign viewers. So we telephoned the N.F.L. and asked where they got their 750 million world-wide viewership. From nowhere at all, it turns out. They just say it! But I wanted to give them the benefit of any doubt, so we telephoned ESPN again and asked how many *possible* foreign viewers there were. Here are the numbers:

| | |
|---|---|
| Possible households in Europe? | 48.40 million |
| Possible households in Latin America? | 3.50 million |
| Possible households in Asia? | 2.00 million |
| Possible households in the Middle East and Africa? | .05 million |
| Total possible households? | 53.95 million |

While not quite complete enough for a totally definitive answer, it is clear where this information leads us. (There are additional telecasters, but their numbers are relatively small.) If we round up the total possible house-holds to 60 million, then for the 659 million non-American viewership number to be accurate, this means not only that every single foreign house-hold (with a television) on the planet was tuned in to the Super Bowl, but that 11 people were clustered around each one of those sets! (Of course, the figures can't actually say that every single household was tuned in, but for every one household *not* watching the game, 22 people in another household must have been!)

How about in the percentage of Americans who think "espresso" is an "overnight delivery system"? Seven turned out to be a teensy bit less honest than the previous example but it's funny enough to forgive the statisticians who managed to evoke that response. I guessed that the data probably arose from a scenario equivalent to the researchers giving people a quiz with a multiple choice of answers—one of which was the purposely hu-morous "overnight delivery system," thus taking innocent advantage of the fact that for a growing percentage of Americans, English is not their primary language. In some Spanish dialects, "expreso" means "special delivery." (And plenty of Americans spell "espresso" as "expresso," complicating the issue.)

So we called Patrice Tanaka and Company, the public relations firm for Krups North America (the coffee company), who commissioned the telephone survey, and it provided us with the details. There had indeed been a multiple-choice quiz, and "an overnight delivery system" was the first choice. This is how the entire question appeared:

---

What is "espresso" (ess-PRESSO)? Is it . . .

. . . an overnight delivery system?

. . . a speedy laxative?

. . . an Italian opera?

. . . a coffee drink?

. . . the latest car from Italy?

---

And the pollsters didn't neglect cappuccino, either!

---

What is "cappuccino" (cappa-CHINO)? Is it . . .

. . . a monk who has taken the vow of silence?

. . . a French actress popular in the 1960s?

. . . an espresso-based drink?

. . . a type of mushroom?

. . . a hillside town in the Italian region of Tuscany?

---

Surely the folks who developed that survey were moonlighting there to supplement their main income as writers for *The Late Show with David Letterman*. (And if not, they should call Letterman right now; maybe they can moonlight for *him*.) Inadvertently funny in retrospect, they were probably just trying to pin down definitely what people knew. And they found out, didn't they?!

Now let's move from the sublime to the ridiculous. This next pair of statistics is an example of the worst kind of *post hoc* fallacy. To be sure, they're only meant to entertain (not to educate), but they do serve as a good illustration of a very common logical error. In sum, we read that since the Berlin Wall fell in 1989, the percentage of *western* Germans who say they have indigestion has gone *up* by 5 percent. We also read that since that same time, the percentage of *eastern* Germans who say they have indigestion has gone *down* by 12 percent. Well, that tickles the American funny bone, all right, but what does it really mean? Nothing. Without a causal relationship, this information is darned near useless. (Now, as for the average level of testosterone in the saliva of male lawyers—well, that sounds pretty straightforward!)

When we look askance at the statistics about the Germans and their political indigestion, it seems almost intuitive to observe that they are empty of meaning. But that's simply because the factors that the statistics try to correlate—in this case, the Berlin wall and stomachaches—appear too far apart in category or classification.

Here's how such statistics can mislead us: The closer the separate factors are in category or class, the more that people will *think* they're causally related, regardless of the apparent evidence (or lack of it). In other words, if the factors have, say, music or medicine or mathematics in common, it has been my observation that our cognitive ability to perceive them as separate entities breaks down. This results in the equivalent of thinking that when Placido Domingo eats an extra piece of cake, Luciano Pavarotti gains weight.

### Facts About Fallacies

The concept of fallacy goes to the heart of intuition (and counterintuition), because a fallacy is not just an error in reasoning; it is a falsity in an argument that appears sound. Perhaps if we began a rigorous study of logic at an earlier age or continued it through a later age, our intuitions would be less riddled with fallacies than they are.

Although there are many more ways to get a wrong answer than there are ways to get a right one, the following are some basic concepts. (Bear in mind that many of the examples demonstrate more than one type of fallacy. For a good exercise in logical thinking, take each example and see how many fallacies it contains.)

*Verbal fallacies arise out of the ambiguity of the words used to express an argument.*

The fallacy of *equivocation* occurs when we use the same word in different senses. To illustrate, here are a few examples:
- "A person can have a germ of an idea. Germs can cause disease. Thus, ideas can cause disease."
- "Death is a subject of utmost gravity. Gravity is what keeps us from falling off the Earth. Thus, death is primarily what keeps us from falling off the Earth."
- "Unattractive people are sometimes called dogs. A dog ages seven years in the same time a human ages only one. Thus, unattractive people sometimes age faster than other people."
- "People often visit psychiatrists when they are down. Down is often stuffed into pillows. Thus, people who go to psychiatrists are often stuffed into pillows."
- "Many bridal gowns have a train attached to their backs. A train can be used to carry freight from one destination to another. Thus, many brides have the ability to carry freight from one destination to another."

Those were obvious, but that isn't always the case. In less-obvious examples, a colleague might communicate a message that we misunderstand and then act upon accordingly. When the error is discovered, we then laugh (or sigh) and say, "Oh, *that's* what you meant!" Or family members might argue when one person reacts negatively to a word that another means to be positive.

The fallacy of *amphiboly* occurs when the sentence construction produces a double meaning. Here are a few examples:
- "I'm having some friends for dinner."
- "The soldiers assigned to the dangerous mission were concerned when they heard Bob's chicken."
- "The anthropologists went to a remote area and took photographs of some native women, but they weren't developed."

The fallacy of *accent* occurs when a false impression is conveyed by an emphasis of the wrong word(s):
- "A snake can only eat green frogs." Does it mean that he can't do anything *else* with his time? Or does it mean that he simply can't *hear* green frogs? Or does it mean that he can't eat *brown* frogs? Or does it mean that he can't eat green *dogs?*

The fallacy of *figure of speech* occurs when we assume that words similar in form are also similar in sense:

- "People find that wool sweaters shrink when they get wet. Thus, sheep become smaller when they get caught in the rain."
- "People often prefer turkey with gravy. Thus, wild turkeys that are covered with gravy attract predators more readily."
- "People usually find dinner plates garnished with sprigs of parsley to be attractive. Thus, people who want to be more attractive should try garnishing themselves with parsley."
- Consider this next one. "You would readily take a prescription drug if you have a physical problem like high blood pressure. Thus, you should readily take a prescription drug if you have an emotional problem." That doesn't sound silly. But it's an elementary logical fallacy. Maybe you *should* take medication for an emotional problem, but not for *that* reason.
- How does this sound? "You would consult a psychiatrist if you have an emotional problem like depression. Thus, you should consult a psychiatrist if you have a physical problem." That's not true, of course. But it's the same logical fallacy. And how does this one sound? "You would readily wear glasses to see better. Thus, you should readily wear glasses to hear better."

The fallacy of *composition* results when we assume that what is true of the parts must be true of the whole. At first glance, this fallacy and the one following (called the fallacy of *division*) may seem to be formal fallacies (instead of verbal ones), but both are considered to be types of equivocation. In the fallacy of composition, the individual terms that comprise a group (i.e., the members of a committee) are equivocally confused with the collective term (in this case, the committee). In the fallacy of division, the opposite is the case. The following are examples of the fallacy of composition:

- "A spider is a beneficial member of an ecosystem. Therefore, introducing millions of spiders into an ecosystem would be advantageous." That seems clearly incorrect to us, but what about this? "One new tax can generate billions of dollars in additional revenue. Therefore, many new taxes could eliminate the budget deficit." It's the same fallacy.
- Or this one? "A minnow is able to evade large predators easily. Therefore, a school of minnows would be able to live safely among them."

That seems clearly incorrect again, but what about this? "A politician can pay off a key figure in his district in return for votes without being detected. Therefore, a politician can pay off many key figures without being detected." On the contrary, with every payoff made, the more likely it is that the unscrupulous politician will be caught.

- Or this one? "A dolphin is a very intelligent animal and even grasps certain crude numerical concepts. Therefore, a school of dolphins together would be able to work through basic mathematics." Clearly incorrect again, but what about this? "Each member of the President's health-care team is very intelligent and understands certain health-care concepts. Therefore, a team of them together would be able to restructure the entire country's health-care system successfully." Maybe it's just wishful thinking, but it's a fallacy, nonetheless.

The fallacy of *division*, then, results when we assume that what is true of the whole must be true of the parts. Here are examples:

- "A group of musicians performs well and is wildly popular. Therefore, if each goes his own way, each will perform well and be wildly popular." (Think of this as the "Beatles Fallacy.")
- "A comedy troupe is funny and hugely successful. Therefore, each member on his own will be funny and hugely successful." (Think of this as the "Monty Python Fallacy.")
- Consider the following: "Swarms of deadly army ants are able to drive people from their homes. Therefore, a single army ant would make everyone run for their lives." Obviously, that was incorrect. But what about the following? "A management team is effective at making sure that the needs of the workers are met. Therefore, as the organization grows, each member of the team can be expected to manage his or her own group of workers successfully." That wasn't so obvious.
- Or this. "A colony of termites can construct large clay and earth residences that even have effective ventilation and a kind of air-conditioning. Therefore, a single termite on its own can still construct a sizable residence for itself." Obviously not.
- But what about this? "A major corporation consisting of several divisions produces goods of high quality and is profitable. Therefore, each division alone can be expected to produce goods of high quality profitably." Not so obvious.

*Formal fallacies arise out of the structure (or "form") used to express an argument.*

The fallacy of *affirming the consequent* results when we assume the "antecedent" (the "if" that comes before) defines the only way in which the "consequent" (the "then" that comes after) can occur. For example:
- "If a person has full-blown AIDS, then his or her T cell count will be low. A person is found to have a low T cell count. Therefore, he or she has AIDS." However, this is not the case. Other conditions do cause low T cell counts.

The twin fallacy of *denying the antecedent* also results when we assume the antecedent defines the only way in which the consequent can occur. For example:
- "If a person has full-blown AIDS, then his or her T cell count will be low. A person is found not to have AIDS. Therefore, he or she does not have a low T cell count." Again, this is not the case, and for the same reason. Other conditions cause low T cell counts.

The last four formal fallacies relate to the "syllogism," a form of deductive reasoning that consists of a major premise, a minor premise, a "middle," and a conclusion. (The "middle" term appears in both the major and minor premises, linking the two together.) Perhaps the most famous syllogism is "Every man is mortal; Socrates is a man; therefore, Socrates is mortal."

Nearly everyone who influences public opinion—from social leaders to politicians—uses syllogisms to arrive at incorrect conclusions, and it's difficult to believe that this is usually by accident. After all, the conclusions nearly always "prove" the point the person wants to make. I'd go so far as to say that never before, in the history of this country, have citizens been so jerked around logically to the point where they have become incapable of making reasonable decisions. This has begun to evidence itself in incredible jury verdicts. By using every logical error known to mankind in an effort to further one or another special interest, we have begun to reap what we have sown—the seeds of intellectual weakness and mental disorder.

The fallacy of *four terms* occurs in a syllogism that uses equivocation in the middle term:

- "One way that plants are different from animals is that some plants bloom in the spring. But some people bloom in the spring, too. Therefore, those people are plants."
- "Trees have wooden trunks that contain important material. But some people have wooden trunks that contain important material. Therefore, those people are trees."

The fallacy of *undistributed middle* occurs in a syllogism when the "middle" term inappropriately links together the two terms in the premises. (A term is said to be "distributed" when it takes into account *all* the members of the set it describes. An "undistributed" term only takes into account a portion of them, making any generalizations invalid.) Here are two examples:

- "Some types of fungi are able to generate their own light and glow in the dark. All fireflies are able to generate their own light and glow in the dark. Therefore, fireflies are a type of fungi." Here, "able to generate their own light and glow in the dark" is the middle term. It is considered undistributed because neither premise states, for example, that "all things that are able to generate their own light and glow in the dark are fungi." The linkage of the major and minor terms (fungi and fireflies) through the middle term is not valid. The major and minor terms simply have a particular feature in common.
- "Truffles are discovered by dogs trained to locate them by scent. Criminals are also discovered by dogs trained to locate them by scent. Therefore, criminals are truffles."

The fallacy of *illicit minor* occurs when the subject of the conclusion is broader than the minor premise allows, as in the following example:

- "Giant sequoias are the largest organisms alive. All giant sequoias are trees. Thus, all trees are the largest organisms alive." In the minor premise, "trees" is undistributed, but it "illicitly" becomes distributed (That is "*all* trees") in the conclusion.

The fallacy of *illicit major* occurs when the predicate of the conclusion is broader than the premise allows, as in this example:

- "Banana trees are actually herbaceous plants instead of trees. All bananas are fruits. Therefore, no fruits grow on trees." In the major premise, "herbaceous plants instead of trees" is undistributed, but, again, it "illicitly" becomes distributed in the conclusion.

*Material fallacies arise out of the fabric (or "material") used to express an argument.*

The fallacy of *ignoratio elenchi* (irrelevant conclusion) occurs in many forms. One is the *argumentum ad hominem,* in which the opponent is attacked instead of the argument. An example is:

- "Darwin's theory of evolution by means of natural selection is wrong because he was a failure at everything else he ever tried."

In *tu quoque,* the opponent is attacked by means of a countercharge, as in:

- "Darwin's theory of evolution is wrong because he didn't publish it until another naturalist proposed a similar idea and was likely to get all the attention for it."

Another form of irrelevant conclusion is the *argumentum ad populum,* in which an appeal to popular passion is invoked, as in:

- "The theory of evolution is nothing but hogwash promulgated by a bunch of godless scientists who want to turn this God-fearing country into another Sodom and Gomorrah."

In the form of *argumentum ad verecundiam,* an appeal to authority is invoked, instead, as in:

- "All the creatures on the Earth were created at one time, exactly as we see them today, because it says so in the Bible."

In the form of *argumentum ad misericordium,* an appeal to pity is invoked, as in:

- "Creation is correct. How can mankind deny its own father? How would *you* feel if you were disavowed by your very own children?"

And in the form of *argumentum ad baculum,* an appeal to force is invoked, as in:

- "Evolution is correct, and if you don't think so, I'm going to break your kneecaps."

This last "appeal to force" sounds almost darkly comical, until we consider how prevalent it actually is. Not only does it work in isolated cases of kidnapping, when the helpless victim is converted by brute force into an adherent of the perpetrator of the crime, whole societies caught in totalitarian nightmares can believe what they *must* believe in order to live without fear or without an intolerable envy of other countries.

One last form of irrelevant conclusion is the *argumentum ad ignorantium,* in which it is maintained merely that because a premise is not known to be untrue, it may indeed be true. For example:

- "There is no proof to support the hypothesis that human beings are descended from barnacles, but neither is there anything in the fossil record that shows this hypothesis to be untrue, either. New discoveries in paleozoology are being made every day, and who knows what the next expedition will turn up?"

The fallacy of *petitio principii* (begging the question) is a circular argument in which the conclusion also appears as an assumption, albeit in different form. Here's a subtle example:

- "The concept of evolution as a long series of chance mutations of various organisms must be incorrect because such a sequence of events would defy the laws of probability." But stating that chance defies the law of probability is a poor argument. That's like stating, "Miracles can't occur because they would defy the laws of nature," also a poor argument. The conclusion appears presumed as a general principle.

A special case of circular argument is *circulus in probando,* also called "vicious circle." In this case, we find arguments like, "The story of the divine creation as told in the Book of Genesis must be true because God would not deceive us," and "The concept of evolution must be true because so many scientists couldn't all be wrong." Neither argument is sound. This is similar to René Descartes, the seventeenth-century French philosopher who is often called the father of modern philosophy, first proving the existence of God by reasoning and then assuring himself that his reasoning was valid by stating that God would not deceive him.

The fallacy of *secundum quid* occurs in two common forms. One is *direct accident*, in which the truth of an abstract principle is inappropriately applied to a specific circumstance. For example:

- "Scientists believe that fish developed lungs and eventually moved out of the oceans and colonized the land. Thus, the fish in our aquarium at home will eventually be found sitting on the sofa."

The other is *converse accident*, which is the reverse. For example:

- "Because the fish in our aquarium at home will never be found sitting on the sofa, no fish ever developed lungs and colonized the land."

The fallacy of *post hoc, ergo propter hoc* ("after this, therefore because of it") is the source of much of the world's superstition. In this fallacy, a particular event is held to be the result of a prior one, rather than an independent, coincidental event. For example:

- "A commencement speaker was struck by lightning while denouncing creationism from an outdoor podium. Therefore, he was punished for denouncing creationism."

In the fallacy of *reductio ad absurdum*, the argument declares that an assumption is false if a contradiction can be drawn from it. Here's an example:

- "Scientists have open minds. Because Darwin was a scientist, he had an open mind. But anyone who categorically denies the authority of the Church has a closed mind. Because Darwin's theory of evolution categorically denied the authority of the Church, Darwin had a closed mind. Therefore, Darwin was not a scientist."

In the fallacy of *plurimum interrogationum*, the argument demands an answer to a question phrased in such a manner that any direct reply supports the implications of the question, as in:

- "Mr. Darwin, have you stopped beating your wife?"

Finally, in the fallacy of *non sequitur*, the conclusion doesn't follow from the argument, as in this example:

- "Because fish have gills and birds have wings, because dinosaurs are extinct and snakes aren't, because the duckbilled platypus and the echidna have characteristics of both reptiles and mammals, because animals need the waste product of plant respiration to survive and plants need the waste product of animal respiration, because plants may need insects for fertilization but earthworms don't even need another earthworm, because whales can sing and dolphins are intelligent, because crustaceans look so much like big bugs and primates look so much like humans, and because nearly every meat on the planet doesn't taste all that much different from chicken, the theory of evolution is correct." That just doesn't follow!

### Fallacies About Facts

The seductive effects of statistics are magnified by their inherently misleading nature, which continually confounds even the most attentive minds; the rest of us just don't have the time even to *try* to separate the fact from the fiction. Without actively (and sometimes with great difficulty) taking the data carefully apart, placing it in appropriate context, and drawing a laborious conclusion from scratch, one will end up with a false impression or the wrong answer. Numerical examples abound. Following is a sampling of "headlines" that are particularly representative of certain types of innocent error ranging all the way to outright mendacity, starting with one that involves that bane of the well-intentioned intuition: averages.

- *People Who Don't See a Doctor for a Common Cold Get Well Faster Than Those Who Do.*

  This statement implies that it's better for us not to visit a physician for a common cold, as though a doctor will somehow make the condition worse. But most of us don't see a doctor for a cold unless we have a complication, and secondary infections extend the illness. So people who visit a physician will typically be among the more seriously ill of the group we're analyzing, and when we average the "sick days" of patients who never needed to see a doctor at all, then compare that figure to the average number of "sick days" for patients who *did* visit a physician, it will show that those of us who saw a doctor took longer to get well.

- *More People Die in Hospitals Than Anywhere Else.*

  No wonder we don't want to go to the hospital. What dangerous places they must be! This statement conjures up visions of unsanitary conditions, negligent nurses, incompetent doctors, and a whole array of frightening accidents, such as mixed-up medications and waking up after routine gall bladder surgery to find one's left leg removed, instead. But the fact of the matter is that you find more sick and injured people in the hospital than anywhere else. (When an emergency medical-service technician extracts you from your crumpled automobile, does he take you *home*?) And so it's likely that more people will die there.

- *More People Die in Bed Than Anywhere Else.*

Elderly folks who've heard this sometimes fear going to sleep, worried they'll never wake up again. But the statement arises out of the understandable fact that when we become ill, we don't play a game of tennis, go out to a restaurant, or wash the car. We go to bed. In addition, we spend nearly a third of our lives there, which means that more people *arise* from bed feeling healthy than anywhere else, too. After all, consider just how few of us sleep on the floor.

- *More Accidents Occur at Home Than Anyplace Else.*

Ergo, it's more dangerous to be at home than it is to be anyplace else. (It's amazing that we feel so comfortable there.) But how can that be? How is it that for me, *my* home is more hazardous, but for my neighbor, *his* home is more hazardous? It sounds true, though; after all, most of my own accidents occurred at home. But there's a good reason for that. Although I've traveled to quite a few places in my life, I've still spent more time at home than at any other single place. So even if it's the safest of all the places I've ever stayed, I'm still likely to have more accidents there.

- *More Violence Is Committed Against Members of One's Own Family Than Against Anyone Else.*

This suggests that it is more dangerous to be around our relatives than to be around strangers. (You'd think relatives would be the *last* people to whom we'd give holiday presents, wouldn't you?) But that's not really the case. The statement comes from the fact that we're around our relatives far more than we're around anyone else, and if one of our relatives is unstable, our risk of having a problem that involves a relative is increased. If, instead, we were all to exchange relatives with a stranger, we would find that our risk of having a problem with *his* relatives would go up.

- *More Automobile Accidents Occur During Rush Hour Than at Any Other Time.*

That implies that it's more dangerous to drive at rush hour. It may be, but not on this evidence! The reason such a large number of automobile accidents occur during those hours is simply because that's when so many people are driving their cars. As another example, just from the sheer volume of the traffic alone, far more windshield wipers are turned on during the rush hours than at any other time. But this doesn't mean that it rains more during those particular times of the day!

- *Per Passenger Mile, Flying Is the Safest Form of Transportation.*

The following doesn't justify the kind of fear of flying that the airlines hope we'll keep in check, but passenger miles are not as flawless a criterion as we might prefer to believe. Flying in an airplane enables us to travel thousands and thousands of miles that we would not otherwise travel, so we can't say that those passenger miles are safer than the armchair traveler's.

Here's another way to look at it: Let's say that we've definitely decided to travel from New York to London. Clearly, we can't travel by car, so we can't compare airline passenger miles to automobile passenger miles. We must compare them to ship passenger miles. Which is safer *for one crossing*? Giving the airplane the benefit of any doubt, let's presume for the moment that they're equal, but that flying allows us to commute back and forth every two weeks for our international business, which we could never do by ship. (We'd transact our business over the phone, or we'd staff two separate offices, or whatever.) So we wouldn't be exposed to all those flights. In short, flying sounds very good by the passenger mile, but when we do fly—even just once around the world—we get exposed to a great many more passenger miles than we would otherwise. We never would have *driven* around the world, would we?

- *The Death Rate Among American Soldiers in the Persian Gulf War Was Lower Than the Death Rate Among Civilians.*

With a certain amusement, many of us took this to mean that the war was so handily won that it was actually safer to be in the conflict than out of it. It's true that this particular war was well run, but you can cite the same false safety statistics about the death rate in the Navy during the Spanish-American War around the turn of the century—a death rate that

Included the sinking of the U.S. battleship *Maine* in Havana Harbor. ("Remember the Maine!") During that time period, the death rate in the Navy was about nine per thousand; the death rate for civilians was about sixteen per thousand.

Recruiters love this sort of specious reasoning. But let's have a closer look at it. Would the death rate of the population go down if we *all* joined the Navy? Of course not. Then, what's going on? Well, the Navy is made up of healthy young men and women, who have a low death rate *wherever* they are, but the general population includes elderly people, infants, and people who are chronically and/or acutely ill. To get to the truth of the matter, we'd need to compare the death rate in the Navy to the death rate of healthy young men and women civilians—not to the entire population.

But of course, you'd think we could trust folks like the California Highway Patrol, the National Highway Traffic Safety Administration, the National Safety Council, and the Drug Enforcement Administration, who inadvertently contributed to the following statistics cited in A. K. Dewdney's *200% of Nothing*.

- 20 percent of fatal traffic accidents are caused by cocaine
- 20 percent are caused by mechanical failure
- 25 percent are caused by marijuana
- 35 percent are caused by falling asleep at the wheel
- 35 percent are caused by suicide
- 50 percent are caused by smoking
- 50 percent are caused by alcohol
- 85 percent are caused by speeding

Dewdney asks, "How are such percentages possible? If the various categories of cause did not overlap at all, one would expect the percentages to add up to no more than 100 percent. But these percentages add up to 320 percent. This number makes sense only if the great majority of the accidents had several separate 'causes,' each of them a major contributor to the fatality." Imagine the percentage total if factors like dangerous surface conditions, poor visibility, carelessness, and plain old human error were added to the total.

Dewdney continues, "The typical accident might well result from the following astonishing combination of causes: the driver first takes some cocaine and marijuana, then drinks a bottle of wine, then heads out to the freeway in a mechanically unsound vehicle, lights up a cigarette, drives at high speed, feels a suicidal urge, and aims for a bridge abutment, falling asleep just before impact. A simpler explanation might be that the agencies reporting these figures were exaggerating. Why? Agencies compete for funding by making their individual mandates seem as important as possible."

My own favorite "bad example" of limited logic comes from an article about cats that appeared in *The New York Times*'s weekly science supplement called "Science Times" on August 22, 1989. It stated, "The experts have also developed startling evidence of the cat's renowned ability to survive, this time in the particular setting of New York City, where cats are prone at this time of year to fall from open windows in tall buildings. Researchers call the phenomenon feline high-rise syndrome."

But the sidebar about how they managed to survive the falls was what interested me the most. It elaborated, ". . . From June 4 through November 4, 1984, for instance, 132 such victims were admitted to the Animal Medical Center. . . . Most of the cats landed on concrete. Most survived. Experts believe they were able to do so because of the laws of physics, superior balance, and what might be called the flying-squirrel tactic. . . .

". . . [Veterinarians] recorded the distance of the fall for 129 of the 132 cats. The falls ranged from 2 to 32 stories. . . . 17 of the cats were put to sleep by their owners, in most cases not because of life-threatening injuries but because the owners said they could not afford medical treatment. Of the remaining 115, 8 died from shock and chest injuries.

"Even more surprising, the longer the fall, the greater the chance of survival. Only one of 22 cats that plunged from above 7 stories died, and there was only one fracture among the 13 that fell more than 9 stories. The cat that fell 32 stories on concrete, Sabrina, suffered [only] a mild lung puncture and a chipped tooth. . . .

". . . Why did cats from higher floors fare better than those on lower ones? One explanation is that the speed of the fall does not increase beyond a certain point, [the veterinarians] said. . . . This point, 'terminal velocity,' is reached relatively quickly in the case of cats. Terminal velocity for a cat is 60 miles per hour; for an adult human, 120 m.p.h. Until a cat reaches terminal velocity, the two speculated, the cat reacts to acceleration by reflexively extending its legs, making it more prone to injury. But after terminal velocity is reached, they said, the cat might relax and stretch its legs out like a flying squirrel, increasing air resistance and helping to distribute the impact more evenly." That seemed to make sense, so I filed the article for future reference.

Some time later, a reader wrote to ask me to "please explain why a cat will land on its feet when it falls from a great height," and I obliged by citing the study in my column, adding the following as an additional point of interest: "Amazingly, the cats that fell longer distances fared better than the others. Of the 22 cats that fell more than 7 stories, 21 survived; of the 13 cats that fell more than 9 stories, all survived. Sabrina, who fell 32 stories onto concrete, suffered only a minor lung puncture and a chipped tooth; I'll bet she was treated to a whole bowlful of tuna that day."

Later, reading these statistics in my published column bothered me, but I didn't know why. It never occurred to me to scrutinize the statements from the original article further. So it wasn't until my assistant dropped a handful of letters about the subject on my desk that I finally took notice. The first was from Pamela Marx in Brooklyn, New York, who wrote, "I have had two cats fall from terraces in two separate instances, and both, unfortunately, died. One was a tenth-floor terrace, and the other was on the fourteenth floor. I never reported these incidents to any medical center and believe that other people probably don't report their cats' deaths, either. You can add my two cats to your list and report that at least two cats died in fifteen falls over nine stories." At that point, the error seemed so obvious that I didn't know how I had missed it in the first place.

▼▼

I'd like to add a cautionary note to your reply regarding cats' ability to land on their feet. You referred to a study reporting the seemingly counter-intuitive result that cats falling from greater heights fare better. As you know, in order to make a generalization to the population of cats falling from windows, one must be sure of having a representative sample of cats from that population. Unfortunately, using the self-selected sample of cats taken to the animal hospital after a fall in no way guarantees that. There must be *many* cats who fall from relatively low heights who appear unharmed to the owners, and so are not taken to the animal hospital. Similarly, there must be a number of cats falling from greater heights who die as a result and receive a shoebox funeral, also never being taken to the animal hospital. This leaves us with a biased sample of cats, not really representing the population at large.

Gregory Hancock
Auburn, Alabama

I didn't know whether to be amused or dismayed. I also don't know quite when I began to pay attention to all the misinformation, disinformation, and flagrant abuse of the general public's lack of education in logic and elementary mathematical skills, but I do know that I found it *everywhere*. I didn't just find a misleading statistic or pronouncement here and there, now and then. I found it (and still do) every day, in every way, throughout the most respected information sources in the country, but most especially from—no surprise—our government. This phenomenon isn't the exception. It's the rule.

# Four

▼

# Why We Lead Ourselves Down the Garden Path

This is how Darrell Huff put it in his book *How To Lie With Statistics.*

▼

The title of this book and some of the things in it might seem to imply that all such operations are the product of intent to deceive. The president of a chapter of the American Statistical Association once called me down for that. Not chicanery much of the time, said he, but incompetence. There may be something in what he says, but I am not certain that one assumption will be less offensive to statisticians than the other. Possibly more important to keep in mind is that the distortion of statistical data and its manipulation to an end are not always the work of professional statisticians. What comes full of virtue from the statistician's desk may find itself twisted, exaggerated, oversimplified, and distorted-through-selection by salesman, public-relations expert, journalist, or advertising copywriter.

If Huff had written his book in the nineties (instead of the fifties), I feel confident he would have put "politician" somewhere on that list. But we'll get to that in the next (and last) section of this book. For now, Huff continues.

▼

But whoever the guilty party may be in any instance, it is hard to grant him the status of blundering innocent. False charts in magazines and newspapers frequently sensationalize by exaggeration, rarely minimize anything. Those who present statistical arguments on behalf of industry are seldom found, in my experience, giving labor or the customer a better break than the facts call for, and often they give him a worse one. When has a union employed a statistical worker so incompetent that he made labor's case out weaker than it was?

As long as the errors remain one-sided, it is not easy to attribute them to bungling or accident.

But first let's consider the reasons for some of the errors that we make *without* outside help.

### Subjective Math and Logic

*Egotistical thinking* is a serious problem. We all want to believe we're right, and so when we form a first impression, even if not in haste, it is difficult to dislodge it. We then proceed to interpret future information so that it confirms what we already think we know.

*Perhaps the desire to be right is even instinctual in the human animal. After all, narcissism has enormous individual survival value, and this may be the mechanism that explains why humans feel such emotional discomfort when making even the most minor of errors.*

*Wishful thinking* is a another significant cause of error and intellectual weakness, but it's more subtle. For example, let's assume we all want a strengthening economy. Now, just for the moment, let's suppose we're Democrats. When a Democrat is in office, we interpret figures as indicating the economy is getting stronger whenever we can. But when a Republican is in office, we interpret figures as indicating the economy is getting weaker wherever possible.

Now let's suppose we're Republicans. When a Republican is in office, we interpret figures as indicating the economy is getting stronger. But when a Democrat is in office, we interpret figures as indicating the economy is getting weaker. There's an intellectual perversity here that we recognize, but, at the same time, deny.

*Counter-intuitive thinking* is the most subtle reason of all for logical errors, which may also make it the most difficult to remedy. The following anecdote from A. K. Dewdney's *200% of Nothing* is a good illustration:

---

A high-school physics teacher once asked his students if they believed in the law of conservation of energy. They did. The teacher took them to a gymnasium where he had suspended a 100-pound iron ball by a cable from a ceiling beam. He invited one of the students to stand on a box some distance from the ominous iron ball. Then he slowly swung the ball [i.e., carried it in an arc by hand] laboriously up to within an inch of the student's nose. He released it and quickly ducked out of the way. The ball swung ponderously away from the student and over the gym floor, gathering speed. "Do you believe in the law of conservation of energy?", the teacher asked, as if repeating a catechism. The student said he did. The ball returned with grand velocity straight at the student's head. With a yowl of terror, he dived from the box just as the ball came to hover, momentarily, an inch or two from where his nose had been, an inch short of where the ball had started its swing.

---

It's very difficult to deny our intuition, which may constitute the denial of an inherited instinct. But even if that's the case, the denial itself has even *more* survival value, creating an even *better* instinct and giving us the greater power inherent in logical thinking. It can and should be practiced.

Remember the uproar over Mattel's "Barbie" when her voice chip chirped, "Math class is tough," back in 1992? Everyone thought it was sexist, and maybe it was. "Ken" probably wouldn't say, "Math class is tough." But the comment would certainly suit "G. I. Joe," and no one thinks ill of his image. Math and logic *are* tough. (We all know that Barbie's not in elementary school; she'd have to be complaining about algebra at the very least.)

Why not admit it? We don't do anyone any favors by cheerfully insisting that math and logic aren't any more difficult than other subjects and that it's merely our attitudes that are the problem. This makes people question their intelligence and privately feel incompetent when it doesn't come easily or they still don't "get it"—something that happens often and that encourages avoidance behavior. And open recognition that math and logic are tough would prepare people to have to work hard at it, budgeting their patience more profitably.

*Passive thinking* takes us back to Huff's comments about the "salesman, public-relations expert, journalist, or advertising copywriter." Human beings are fairly passive animals. Cultures that developed in areas where the climate was temperate and food was plentiful exerted themselves less in all ways than cultures that developed in more demanding conditions. You'd think that peoples who had the least difficult circumstances should have developed the most advanced systems of reading, writing, and arithmetic, but that wasn't the case. It was just the opposite.

*Active thinking* requires strong powers of logic, which readily can be taught in three parts, adaptable to our present system of education. (This book attempts a rough approximation of the approach, although not in consecutive order.)

First, we should study mathematics throughout our early years, including elementary and high school. Math is, after all, just logic with numbers. (However, while the later study of mathematics is fine, of course, I believe it has little impact on continuing logical development because it remains narrow to numbers.) Second, in high school and college, we should study logical processes with particular attention to the fallacies, which could use great expansion. Strong powers of logic help us achieve our goals in life, and in our increasingly complex society, opportunities for success expand enormously for those with the ability to reason well.

Third, in college and beyond, we should study practical applications of the art of reasoning, especially where it has political and economic ramifications. This is easiest when we take examples from real life—politicians use false argument more than anyone else, so they appear prominently in this book—and subject them to the power of the human mind at work. Because Americans have unprecedented influence in directly determining the destiny of our country, it is becoming increasingly critical that we wield it wisely.

We humans tend to believe what we're told unless we take the trouble to find out otherwise, and few of us have the inclination, the time, or even the means to accomplish such an investigation. This is one of the most serious weaknesses inherent in a democracy that values "one man (or woman), one vote." We are prone to elect people who say what we want to hear, thereby putting in leadership positions more people who seek votes simply as a means to power, and fewer people of great stature. And perhaps the greatest irony of all, then, is that with this system, the people who have the least economic success have the most influence on our economy.

## About the Appendix

We should give more attention to studying the mechanics of how people go wrong in their thinking. Not long after the Monty Hall episode had gone down from a full boil to a continuing low-level simmer, we received a letter from Donald Granberg, a Professor of Sociology and a Research Associate at the Center for Research in Social Behavior at the University of Missouri in Columbia. Dr. Granberg, the coauthor and coeditor of two books and the author of some eighty journal articles and chapters, wanted to study the letters we had received about the game-show columns, and so we happily shipped him a couple of dozen cubic feet of mail. With the help of just about everyone he could find, he completed a thorough examination of it, the result of which appears as an appendix to this book. I recommend reading it.

# Five

▼

## How Even Our Health Is Affected

Suppose you have a serious health problem with a mortality rate of 10 percent: Ninety out of a hundred people with this problem will suffer no consequence, but ten will die from it soon. A physician offers you a surgical procedure that will cure the problem if successful, but the procedure itself is risky and would kill 3 percent of those same one hundred people. Do you want to try the surgery?

Now let's say you have that same problem with the same mortality rate of 10 percent. But this time your physician offers you a different surgical procedure, one that is totally safe, never kills anyone, and reduces the 10 percent mortality rate to 4 percent. Do you want this surgery?

If you'd want the second surgery, but not the first one, you have plenty of company. With the first surgery, many people will say that because there's a 90 percent chance that they would be in the lucky group that lives, they'd rather not try a risky surgical procedure that itself would kill 3 percent of the people with that health problem. But with the second surgery, people will say that they certainly *would* want a procedure that is totally safe and reduces the mortality rate dramatically, from 10 percent down to 4 percent.

However, if everything else is equal, the first surgery is the better bet. If all of the people with the health problem were to have the procedure, 97 percent of them would be cured, and the surgery itself would kill 3 percent. But if all the people were to have the second surgery, 96 percent would suffer no consequence, and the illness would kill 4 percent.

Now that we're warmed up for this, let's look at another counterintuitive problem.

## Drug Testing and AIDS Testing

The following exchange was published in the "Ask Marilyn" column:

▼▼

Dear Marilyn:

A particularly interesting and important question today is that of testing for drugs. Suppose it is assumed that about 5 percent of the general population uses drugs. You employ a test that is 95 percent accurate, which we'll say means that if the individual is a user, the test will be positive 95 percent of the time, and if the individual is a nonuser, the test will be negative 95 percent of the time. A person is selected at random and given the test. It's positive. What does such a result suggest? Would you conclude that the individual is highly likely to be a drug-user?

Charles Feinstein, Ph.D.
Santa Clara University

Dear Charles:

Given your conditions, once the person has tested positive, you may as well flip a coin to determine whether s/he's a drug-user. The chances are only fifty-fifty. (The assumptions, the makeup of the test group, and the true accuracy of the tests themselves are additional considerations.) This is just the sort of common misunderstanding that should give great pause to those who will make the decisions about testing.

Shortly afterward, we received this next letter. The reader didn't ask to remain anonymous, but we thought it best anyway:

▼▼

Dear Marilyn:

I read your column on random drug tests, and it hit home very hard. I am a truck driver. I went for a random drug test and showed a false positive. This has cost me my job, and I am now in a drug-rehabilitation program. Neither my boss, the union, the clinic that gave the test, nor the Department of Transportation will even consider the possibility of a false positive. I desperately need to know more to try to save my job and clear my record.

T. M.
Chicago, Illinois


As Professors of Statistics, we found your response to the drug-testing question perplexing and, indeed, incorrect.

Another way of framing the question is, "Out of all the people who test positive, what percentage are actual users?" On the first test, the correct response is 95 percent, meaning that the test is incorrect only five percent of the time. This is *not* a fifty-fifty proposition. We would also argue that the test should be given twice. The likelihood of error on two consecutive tests is only twenty-five times out of ten thousand. We hope you will re-address this.

Paul A. Susen, Ph.D.
Herman Gelbwasser, Ph.D.
Mount Wachusett Community College


What we should be concerned about is the probability of a nonuser being wrongly classified, and this probability remains at 5 percent. A second drug test would decrease the probability of misclassifying a nonuser to ¼ of 1 percent. We'll take the risk.

Eric J. Villavaso, Ph.D.
Gerald H. McKibbin, Ph.D.
U.S. Department of Agriculture
Mississippi State, Mississippi


Next time you respond to such a question, I think you should keep your political views to yourself.

Steve Kovler
Monroe, New York


Dear Readers:

Urging understanding among those who will make the decisions about testing isn't a political statement. This is a complex and important issue; study is required. Drug-testing is a powerful tool, and like all powerful tools, it must be handled with care. Simply eating poppy-seed pastry can make a person test positive for morphine. So will a cough syrup containing codeine. Even over-the-counter drugs cause a similar problem.

The original "fifty-fifty" answer is correct. Also, it often doesn't help to repeat the same test, because the false reports aren't random. Instead, they're more likely to come from analytic sensitivity (the proportion of positive results in actual positive samples) and analytic specificity (the proportion of negative results in actual negative samples), and these are determined by biochemical factors, not statistical ones. That is, we're not referring to "lab error."

Here's how the "fifty-fifty" answer is determined. Suppose the general population consists of 10,000 people. Of those people, we assume for this problem that 95 percent of them (9,500) are nonusers and that 5 percent of them (500) are users.

Of the 9,500 non users, 95 percent of them (9,025) will test negative. That means 5 percent of them (475) will test positive. Of the 500 users, 95 percent of them (475) will test positive. That means 5 percent of them (25) will test negative. These are the totals:

| 9,025 | true | negatives | (nonusers) |
|---|---|---|---|
| 475 | false | positives | (nonusers) |
| 475 | true | positives | (users) |
| 25 | false | negatives | (users) |
| 10,000 | total | population | |

There are 475 "false positives" and 475 "true positives," a total of 950 positives, so when we find an individual in that positive group, there's only a fifty-fifty chance s/he's a user.

But let's suppose instead that a randomly chosen person tests negative. From the above calculation, we can see that there are 25 "false negatives" and 9,025 "true negatives"—a total of 9,050 negatives—so for an individual in that *negative* group, there's an overwhelming chance (more than 99 percent) that s/he is *not* a user.

Here's another letter:

Dear Marilyn:

Your answer regarding drug-testing is correct. Drug-testing should not be done in this fashion because the results of testing may affect employment, liability, etc. There are about 105 laboratories certified by the Forensic Urine Drug Testing Program and/or the National Institute of Drug Abuse. In these laboratories, the specimens that test positive in the screening procedure are retested for confirmation by an even more specific method, raising the predictive value of a positive test.

It is vitally important that your readers understand that the situation described by your reader may exist if testing is done outside of a certified laboratory, such as an employer or a noncertified laboratory. This situation does not occur if testing is done by a certified laboratory.

David A. Mulkey, M.D.
W. Howard Hoffman, M.D.
David A. Miller, M.D.
Peter A. Scully, M.D.
Desert West Drug Testing Consultants
Las Vegas, Nevada

Dear Readers:

Suppose the overall performance goes all the way up to 99 percent, instead. Is *that* good enough? No!

The Centers for Disease Control (CDC) state that the two tests (enzyme immunoassays or "EIA" and Western blots or "WB") combined have better than a 99 percent overall analytic performance accuracy rate in identifying human immunodeficiency virus type 1 (HIV-1), but only if they're taken repeatedly. (The actual rate is unknown.) This rate is the percentage of correct test results in all specimens tested. With a 99 percent rate, if a population of 10,000 were tested, 9,900 would receive correct results, but 100 would receive erroneous results, either false positives or false negatives, including indeterminates.

The CDC also state that of the errors, they have no data on how many are false positives versus false negatives. However, if we use 99 percent as an example, the false positives would have to be less than 4/10s of the erroneous results because the CDC estimate that .4 percent of Americans are "HIV positive." That is, if false positives accounted for fully 4/10s of the errors, then the .4 percent of people who are HIV positive would all be false positives, and we know that's not the case.

Let's try assuming that false positives are only 2/10s of those errors, leaving false negatives accounting for the remaining 8/10s. So, of those same 100 people with erroneous results, 20 percent would have false positives and 80 percent would have false negatives. While those "false negative" people would be an unwitting threat to sex partners, at least most people are aware that "negative" doesn't mean "safe."

But there's another ramification: The CDC estimate that .4 percent of Americans are "HIV positive." In a population of 10,000, that's 40. But that number must include all the false positives, which we assumed to be twenty, leaving only 20 people actually infected. This leads to an interesting conclusion.

In this 99 percent scenario, there are as many "false positives" as there are "true positives." Even if both the AIDS test results (EIA and WB) are positive, the chances are only fifty-fifty that this individual is actually infected. That's why people with HIV-positive results must be sure to get tested repeatedly over the following months. The error rate is high with only two tests. (The CDC's Morbidity and Mortality Weekly Report shows an overall performance rate of only 98.4 percent on the Western blot alone, far worse than our example.) Even after years, you may not fall ill. You may not have AIDS at all. You may just be a "positive-tester."

The implications are broad for people tested at random. For example, I was tested during a routine insurance exam, but because I have no risk factors, I had no concern about having AIDS. However, I was concerned about a "false positive," for which everyone has a risk. (In the 99 percent example, that risk is 20 out of 10,000.) Such a result might have made me uninsurable and destroyed both my personal and professional life. Fortunately, my test was negative. Anyone, however, might be one of those unlucky false positives, and that's the very serious risk of testing.

There's also the danger that "false positive" people will feel they don't need to avoid sex with other "HIV positive" people—a good route to infection that the "false positive" people *don't* already have (the way they think they do).

But there's room for optimism in these statistics. An individual who is a random "positive" can find hope in them. This doesn't mean that he or she can take chances with other people's lives, of course, so each person must behave as though he or she is actually infected. But inwardly, the random HIV positive individual has reason to be cautiously optimistic.


Reaction to this column was mixed. On one hand, one reader (Howard Laitin, Ph.D., Hughes Aircraft) wrote, "As part of my Ph.D. dissertation research (completed at Harvard University in 1956), I reviewed the then-existing accuracy (and the potential accuracy) of a wide range of medical diagnostic tests. I also modeled the accuracy that would be required . . . before various classes of proposed tests could actually serve as reasonable, cost-effective aids to diagnosis. Your column is the first publication that I have seen (in nearly forty years of looking) that provides a simple and accurate explanation. Congratulations. I hope that your column helps educate the public and that it also helps educate medical researchers and medical-care practitioners."

Another reader (George Dellaportas, M.D.) wrote, "I am glad that you brought this issue into [public view]. It is also important in medicine, because all diagnostic tests and procedures have sensitivities and specificities less than unity. Therefore, their results may or may not predict the condition, depending on its prevalence. As a matter of fact, because these tests are first tried on an affected-with-the-condition population sample of high prevalence, they tend to be quite accurate and enthusiastically launched. However, when tried in the general population, they produce a lot of false positive—and quite unhappy—persons, as in the case of your drug-tested [trucker]."

But the Centers for Disease Control were not at all pleased. This next reader (K. Barton Farris, M.D., a pathologist) echoed their concern. "I read your column concerning drug and AIDS tests, and the explanation you gave is absolutely correct. However, I am concerned that your readers may now have a lack of confidence in lab tests.

". . . Positive predictive value depends not only on the characteristics of the test system (i.e., test sensitivity and specificity), but on the prevalence of the disease for which the test is done. Prevalence is the number of cases of a disease at a given time in a given location. As prevalence increases, the number of true positives increases and the number of false positives decreases, thus increasing the positive predictive value and the usefulness of a positive test result.

"I teach this concept to . . . medical students each year. . . . The point that I stress, and that I would like your readers to remember, is that when a physician performs an adequate history and physical exam and establishes a differential diagnosis list, he/she has effectively increased prevalence, making any test result more meaningful. It is only when tests are performed at random . . . that there is a problem with predictive value."

I caution readers, however, not to see the foregoing as a medical statement.* Instead, we used this scenario to illustrate an intriguing mathematical/logical phenomenon. Neither, however, should the reader construe this phenomenon as peculiar to drug-testing or AIDS-testing or even think it is particularly unusual. Rather, counter-intuitive results lie hidden throughout our lives, confounding both citizen and scientist daily.

### Screening for Breast Cancer

Scott Plous, Ph.D., an Assistant Professor of Psychology at Wesleyan University who publishes widely and has been the recipient of many awards, including a MacArthur Foundation Fellowship in International Peace and Cooperation, wrote to tell us of a fascinating case study:

▼▼

Dear Ms. vos Savant:

In your recent column on the misinterpretation of AIDS- and drug-testing, you made the point that "the problem is not a hypothetical one." How right you are! In my book, I discuss a study in which physicians were given a problem similar to the one in your column. They were asked what the odds of breast cancer would be if a woman who was initially thought to have a 1 percent risk of cancer ended up with a positive mammogram result (a mammogram accurately classifies roughly 80 percent of cancerous tumors and 90 percent of benign tumors). Even though the correct answer is around 8 percent, ninety-five out of a hundred physicians who were asked this question estimated the probability of cancer to be

*In the interest of updating the medical situation, it should be noted that the Food and Drug Administration has since approved an immunofluorescence assay that has had good success in confirming cases that are indeterminate using the Western blot assay. It can also be used as an alternative to the Western blot and the enzyme immunoassays. However, these additional test results also require cautious examination and interpretation.

about 75 percent. Misinterpretations such as this, if left uncorrected, could lead physicians to recommend unnecessary mastectomies.

Scott Plous, Ph.D.
Wesleyan University

At this point, careful readers may be looking back and saying to themselves, "8 percent? Could that be a typo? Should it be 80 percent?" Plous explains it in his book *The Psychology of Judgment and Decision Making*, which won the William James Book Award from the American Psychological Association.

▼

Suppose you are a physician who has just examined a woman for breast cancer. The woman has a lump in her breast, but based on many years of experience, you estimate the odds of a malignancy as one in a hundred. Just to be safe, though, you order a mammogram. A mammogram is an X-ray test that accurately classifies roughly 80 percent of malignant tumors and 90 percent of benign tumors. The test report comes back, and much to your surprise, the consulting radiologist believes that the breast mass is malignant.

*Question:* Given your prior view that the chances of a malignancy were only 1 percent, and given test results that are 80 or 90 percent reliable, what would you say the overall chances of a malignancy are now?

According to David Eddy . . . , ninety-five out of one hundred physicians who were asked this question estimated the probability of cancer given a positive test to be about 75 percent. In reality, though, the . . . correct answer is only 7 or 8 percent—a tenth of the typical answer found by Eddy. Apparently, physicians assumed that the chances of *cancer* given a *positive test result* were roughly equal to the chances of a *positive test result* given *cancer*. Decision researcher Robyn Dawes calls this mistake "confusion of the inverse."

(To see why the correct answer should be only 7 to 8

percent, it is necessary to understand a fairly subtle rule of probability called "Bayes' theorem.")*

These results come as a surprise to many people, but what is equally surprising is the way that the physicians in Eddy's study reacted when they were informed of their mistake. In the words of Eddy, "The erring physicians usually report that they assumed that the probability of cancer given that the patient has a positive X-ray . . . was approximately equal to the probability of a positive X-ray in a patient with cancer. . . . The latter probability is the one measured in clinical research programs and is very familiar, but it is the former probability that is needed for clinical decision making. It seems that many if not most physicians confuse the two."

Although confusion of the inverse is by no means limited to physicians, there are very few areas in which it is more

---

*According to Bayes' theorem, the correct way to estimate the odds of cancer given positive test results is as follows: $p(\text{cancer}|\text{positive}) =$

$$\frac{p(\text{positive}|\text{cancer})p(\text{cancer})}{p(\text{positive}|\text{cancer})p(\text{cancer}) + p(\text{positive}|\text{benign})p(\text{benign})}$$

The way to read "$p(\text{cancer})$" is "the probability of cancer," and the way to read "$p(\text{cancer}|\text{positive})$" is "the probability of cancer *given* that the test results are positive." The former quantity is a *simple* probability, and the latter is a *conditional* probability.

The probability on the left side of the equation is the quantity Eddy asked physicians to estimate, and the quantities on the right side of the equation were given in the problem as:

| | |
|---|---|
| $p(\text{cancer}) = .01$ | (the original estimate of a 1 percent chance of cancer) |
| $p(\text{benign}) = .99$ | (the chances of not having cancer) |
| $p(\text{positive}|\text{cancer}) = .80$ | (an 80 percent chanced of a positive test result given cancer) |
| $p(\text{positive}|\text{benign}) = .10$ | (a 10 percent chance of falsely identifying a benign tumor as malignant) |

Once these values are known, it is a simple matter to plug them into the equation and calculate the conditional probability of cancer given a positive test result: $p(\text{cancer}|\text{positive}) =$

$$\frac{(.80)(.01)}{(.80)(.01) + (.10)(.99)} = \frac{.008}{.107} = .075 = 7.5 \text{ percent}$$

important than medical diagnoses. In the event that life and death decisions must be made, there is little solace in answering the wrong question correctly.

What should you do in such a situation if using Bayes' theorem is impractical? The answer is to pay close attention to what statisticians call the "prior probability." The prior probability is the best probability estimate of an event before a new piece of information (e.g., a mammogram result) is known. In the breast cancer problem, the prior probability is the original estimate of a 1 percent chance of cancer. Because the prior probability is extremely low and the mammogram test is only 80 to 90 percent reliable, the postmammogram estimate should not be much higher than the prior probability. The same principle is true with virtually any outcome that is initially thought to be extremely improbable (or probable) and is "updated" with a somewhat unreliable piece of information. According to normative rules such as Bayes' theorem, the absolute difference between the revised probability and the prior probability should not be large.

Again, understand that we're not making a medical statement; instead, it's a mathematical one. Nevertheless, when we begin to see the vast extent to which a command of math and logic is relevant to our everyday lives, it can be disconcerting, to say the least. After all, what percentage of the active voters in this country come equipped with a strong natural logical inclination or have a solid educational background in logic?

## Questions We Should Ask

- Did the presenter of the information generate the numbers himself or herself?
- If not, who or what was the original source?
- Is that source a reputable one, like a university, laboratory, or medical association?
- If so, did the source generate the interpretation or conclusion, too? Or just the raw numbers?

- Does the presenter of the information have a professional agenda he or she wants or needs to support? Or a personal one?
- Is there a conscious bias? (Every newspaper's front page is biased; the notion that a newspaper's position only comes through on its editorial position is no longer even quaint; it's foolish.)
- Is there an unconscious bias?

- Are the numbers presented in context?
- If so, is the context truly relevant?
- Are the numbers presented in comparison?
- If a comparison, are the units of measurement truly equivalent?
- Is the term "average" used?
- If so, is it stated whether it's a mean, median, or mode? Are the upper and lower limits defined?

- Would having more knowledge about the subject affect your understanding?
- Would knowing comparable figures for other time periods affect your understanding?
- Would knowing comparable figures for other locations affect your understanding?

- Has favorable data been stressed and unfavorable data hidden?
- How do you know that?
- Has an unfavorable term been replaced by one that merely sounds better?
- Can you use the same numbers to say something else entirely?

- Is an index presented without its base?
- Are percentages presented without the actual numbers?
- If a number is said to have risen or fallen by a certain percent, from what number did it rise or fall?
- If a number illustrates an increase or decrease over time, have experts learned more about the subject (also over time) that would change what would be included in those figures?
- If comparisons are being made over time, has the definition changed to become more or less inclusive?
- Is the number of cases reported likely to be all of them?

- Could there be a causal factor omitted in order to imply that some other factor was the cause of change?
- If a number (or change in a number) is correlated with another number, what is the evidence that one is actually responsible for the other?

- Can past or present figures accurately predict future figures? Or are they inherently self-limiting?
- For what purpose was a particular study done?
- Do the results seem to agree with what that particular organization would have liked to prove?
- If people were surveyed, would they be afraid of making a politically incorrect statement to a stranger at the door, on the phone, or in the mail?

It's becoming increasingly clear that *every* number must be viewed with caution.

# Part Three

▼

HOW POLITICIANS EXPLOIT
OUR INNOCENCE

# Six

## The Election of 1992
## (The Advent of "The Reign of Error")

The year 1992 was a major election year, and the presidential campaigns focused on the economy. For every U.S. citizen, the stakes were high. In the overview of the national survey of American economic literacy conducted jointly that year by the National Center for Research in Economic Education and the Gallup Organization (discussed in Part One of this book), it was noted that ". . . for most people, the knowledge base for understanding or discussing most economic issues is inadequate. *This economic illiteracy has the potential to misshape public opinion on economic issues and lead to economic policies that have negative or perverse effects on the economy.*"

The survey states that "The American public, high school seniors, and college seniors show widespread ignorance of basic economics that is necessary for understanding economic events and changes in the national economy. When asked questions about fundamental economics, only 35 percent of high school seniors, 39 percent of the general public, and 51 percent of college seniors gave correct answers." For example, "Only 22 percent knew the current national rate of employment; of those who gave a response, most thought the rate was much higher than it was." And, "Only 11 percent knew the current national rate of inflation."

And who knew what? Many people will be either delighted or dismayed to discover that among the more interesting findings were the following:

- "In general, males scored higher than did females within each of the groups surveyed. Receiving the highest average scores were male college seniors. High school senior females and general public females received the lowest scores."

- "Political conservatives tended to receive higher economic knowledge scores than did middle-of-the-road or liberal respondents. This was the case in each of the groups included in the survey."
- "Economic knowledge was also found to be related to respondent's level of income. Wealthier respondents received higher average scores in terms of economic knowledge than did persons with lower incomes."
- "The current projections call for a $400 billion deficit in the Federal budget for the 1992 fiscal year (October 1, 1991 to September 30, 1992). About two-thirds of the respondents in each group thought the deficit would be $700 billion or $1 trillion. In fact, 43 percent of the general public, 43 percent of the high school seniors, and 48 percent of college seniors thought the federal deficit would be about $1 trillion."
- "Profits as a percentage rate return on investment have averaged about 13 percent for most major American corporations over the past decade. On the average, the general public thought that American corporations made approximately 32 percent profit as a return on investment."

Not only were people quite wrong, they were quite wrong all in the same direction: They thought that things were much worse than they were. The 1992 presidential campaign began in the shadow of a recent recession, and the fact that it was already over had not yet registered in most American minds. Given this lack of awareness, it is understandable that personal finances and concerns about the future of the economy dominated the national political agenda and were passionately exploited by those who stood a chance to gain from the situation.

And so, two hundred years after the period of the French Revolution known as "The Reign of Terror," along came the advent of the period in modern politics I call "The Reign of Error." Both of the challengers to President George Bush—Governor Bill Clinton of Arkansas and H. Ross Perot, a billionaire businessman from Texas—published books that focused on their proposals to reduce the deficits and to rejuvenate the economy after what they maintained had been a decade of neglect and destruction by Bush and his predecessor, former President Ronald Reagan.

Never before had attempts to educate and/or exploit American voters with information, misinformation, and disinformation about the economy played such an important role in a presidential election. Perot even took to buying expensive blocks of national television time to present his charts and graphs and simple solutions, thus guaranteeing himself a place in America's popular political culture. And from all the candidates, numbers about job growth, income inequality, deficit spending, and just about every other economic matter rained down upon the prospective voter.

Was all this an effort at enlightenment? Or was it an unvarnished attempt to exploit Americans' numerical naïvete to achieve political power? Surely, both. But what percentage was educating and what percentage was manipulative is what counts, and those percentages don't appear flattering to the candidates. When numbers are used by politicians, in particular, they often don't mean what they seem to mean. The realities about the economy—and especially the numbers generated by the government to explain them—are not as clear and simple to express as the candidates made them out to be. Both their interpretations of recent economic history and their visions of the future are blurred by illegitimate assumptions, lapses in logic, outright misstatements, and inherent measurement uncertainties.

In an effort to accomplish what the candidates failed to do (and probably never intended to do, anyway), let's take a closer look at the 1992 campaign—destined to become a classic example of voter manipulation—and see how the candidates used numbers to their advantage. In the process, we can learn how to cope better with elections of the future. When politicians talk numbers, they're talking economics, and when we combine American number blindness and lack of economic knowledge, we have a recipe for electoral disaster.

To concentrate energy on the most popular and/or influential publications, 1992 pre-election news clips and commentary were gathered from the following sources for review:

> *Putting People First* by Bill Clinton and Al Gore
> *United We Stand* by Ross Perot
> *The New York Times*
> *The Wall Street Journal*
> *Time*
> *Newsweek*
> *U.S. News and World Report*
> *Business Week*

Most of the newspaper and magazine clips are organized in chronological order so the reader can get a better sense of events as they happened; the excerpts from the campaign books were placed where they best fit the context. The first opens this section because it explains so many concepts important to the understanding of the rest.

## The Worst Economic Record in Fifty Years?

*"The Republicans in Washington have compiled the worst economic record in fifty years: the slowest economic growth, slowest job growth, and slowest income growth since the Great Depression."*

> *Putting People First*
> by Bill Clinton and Al Gore
> 1992

While on the campaign trail, our favorite baby boomers based much of their rhetorical appeal on their portrayal of the economy as an increasingly fragile structure that had been all but looted by the middle and upper economic classes while in the control of the Republican Party during the preceding twelve years. They also characterized those years as having been a period of private economic pain for the bulk of the population who were nearer to the bottom of that economic pyramid. The statistic cited above from Clinton and Gore's campaign document makes a good example of that tactic. But does it also make good logical and mathematical sense?

First, much of the state of the economy was their own party's doing. Although Republicans had won the presidency at times, Americans hadn't had a Republican government in forty years. The Democrats had controlled both houses of Congress since the 1930s, with the lone exception of Eisenhower's first term in the 1950s. The Republicans gained control of the Senate only in the 1980s during the Reagan Administration but were unable to hold it throughout his second term.

In our political system, it is Congress that has the only real authority to generate and pass budgetary and tax policies—the most powerful ways in which the federal government affects both the economy and the important numbers in our private lives. Though the President submits suggested budgets, in the end he must merely accept or reject the budgets that Congress offers back to him. During the Reagan presidency, for ex-

ample, actual government spending exceeded the spending requested by
Reagan in every year but 1984. In one case, the additional spending
reached $50 billion.

Politicians specialize in statements such as that of Clinton's and Gore's,
which is linguistically structured to imply a numerical specificity that, in
fact, is absent. Growth is a concept that must be measured from one time
period to another, and Clinton and Gore identify the period they used as
a standard of comparison: the preceding dozen years of Republicans holding the highest office in the Executive branch of government. But in doing
this, they combine the Reagan and Bush years, which is misleading. The
administrations of these two Presidents did not combine to represent one
approach to matters of economic policy and its effects on growth, jobs,
and income. For example, average annual real domestic spending under
Reagan increased just 1.0 percent, but under Bush, it increased 8.7 percent, a significant difference. Moreover, when assessing blame and/or
credit, we must add the factor that there were also different Democratic
Congresses operating throughout these periods—each of them pursuing
their own particular course for their own political reasons that varied
markedly from the two different Republicans who were President at the
time.

But even if we ignore the different policies pursued (and results obtained) by the two different Presidents and the various Congresses in
power, we still have a problem: We don't know to which other specific
time period this dozen years is to be compared. Clinton and Gore refer
to the "slowest . . . growth since the Great Depression," but they don't
specify any of the many possible periods within this period over which
this growth can be compared. This makes a great deal of difference.

There are other gaps in specificity that make checking the numerical
realities tricky. The phrase "economic growth" in an overall sense can be
defined in many ways; the most popular measure today is Gross Domestic
Product (GDP). Gross Domestic Product only recently (in December
1991) replaced the older measure, Gross National Product (GNP), though
the difference between these two huge macroeconomic variables is fairly
minor. (Technically, GDP attempts to measure the production that occurs
within the U.S. border, regardless of the nationality of the workers and
the owners of capital, while GNP attempted to measure the production
of U.S. workers and companies regardless of where they lived or worked.
In 1991, the difference between the two measures was only $13 billion,

big money to the average citizen, but relatively small money to the government—it represented only 0.2 percent of the GDP. A government official once said, "A billion here, a billion there, and pretty soon you're talking about real money.")

To arrive at GDP figures—that is, the total production of both goods and services in the United States economy—the Commerce Department cobbles together large numbers of reports from other governmental agencies, as well as reports from private companies and individuals, then compiles them into a broad set of statistics on national income and production. The primary understanding, then, should be that these economic aggregates are not projections of national prosperity or well-being in any sense. They are simply dollar figures, and dollar figures out of context are meaningless. Take personal income, for example. Without knowing the price of food, clothing, and shelter, and without knowing the tax rates and other relevant numbers, an income number tells us nothing about the financial comfort (or discomfort) of its earner.

Even when it's in context, GDP may still be out of focus. Crudeness is inherent in the act of assembling macroeconomic numbers. Balance sheets mean very different things from one individual to another and from one company to another. Wealthy individuals and enormous companies can topple overnight, and they do so routinely. There are many more factors to consider.

For example, let's suppose that you routinely do your own housecleaning and cooking, and so does your neighbor. The GDP is unaffected. Now let's suppose you get too busy to do your own housecleaning, so you hire your neighbor to do it. The GDP goes up. But now your neighbor is too busy to do his own cooking, so he hires you to do it. The GDP goes up even more. Are these two households better off? Maybe. Otherwise, presumably, you wouldn't be doing it. But in what sense can "national production" be said to have increased? Using the GDP as a standard, we could make our country two or three times as "productive" simply by hiring each other to perform reciprocal services. And as this would also be an income-tax boon, the deficit could be reduced. With the GDP up and the deficit down, the country would be booming. But in fact, the country wouldn't be more productive at all, and we'd all be poorer by the amount we paid to the Internal Revenue Service.

Which brings us to what the concept of employment is all about—people hire each other. This highlights the weakness of trying to measure slippery concepts like "economic growth" in numbers. The fact remains, however, that when the numbers are "good," people feel better—regardless of whether they are indeed better off—and when the numbers are "bad," people feel that way, too.

But let's take the numbers at face value and examine GDP rates as they shed light on Clinton and Gore's statement. For the purpose of analyzing the point they made—that disparate economic policies have made a difference in the health of the American economy—it seems reasonable to disengage the Bush years from the Reagan years. (After all, Clinton was running against Bush, not Reagan.) During the Reagan years from 1981 to 1989, the average annual growth in GDP per capita (that is, divided evenly among the entire population) was 1.99 percent. The Bush years were less expansive. From 1989 to the second quarter of 1993, the average annual growth in GDP per capita was only 0.33 percent.

Looking just at Reagan, then, it's not hard to find at least one time period since the Great Depression in which average annual growth in GDP per capita has been slower. Assuming that we consider the time periods defined by presidential terms of office, the Carter years of 1977 to 1981 produced only a 1.03 percent rise, and the Nixon years of 1969 to 1974 produced only a 1.39 percent rise.

It's also not hard to find a period in which average annual growth in GDP per capita was faster. The Johnson years of 1963 to 1969 produced a rise of 2.39 percent. So without a time period for comparison being specified, it's also not hard to see whatever you want to see by handpicking your own standard period to make the point you want to make.

Another measure of economic growth is the average annual growth in real private net worth, including both physical possessions and financial wealth. This, assuming an adequate method of assessment, may be more meaningful to the well-being of individual citizens than GDP/GNP measures. On those terms, the Reagan presidency, with its 3.9 percent growth rate, beats out the Carter era with its 3.6 percent and far outshines the Kennedy years with their 1.8 percent and the Nixon years with their 0.4 percent. On Bush's watch, this measure showed a dismal growth of 0.06 percent, underscoring the misleading nature of fusing the Republican policies of the twelve Reagan-Bush years.

What about job growth? Between 1980 and 1990, employment increased by about nineteen million jobs. But let's look at how these figures are obtained, so we can decide how much importance to give them. Employment data is collected by the Bureau of Labor Statistics (BLS) through several methods. One is a survey of sixty thousand households done by the Census Bureau, which asks people questions about the work activities of all householders over the age of sixteen. As the people at Nielsen do, they extrapolate those numbers (which represent perhaps a thousandth of the actual population) to make a statement about the nation as a whole. The standard error for this sample is 0.11, which is supposed to mean that there is a one-in-three chance that the actual figure will be off by more than 0.11 percent, so any change of less than 0.2 percent is considered statistically insignificant.

The BLS also surveys employers, asking them questions about their numbers of workers, types of workers, average weekly hours, average hourly earnings, and the like. The BLS proudly claims that "the sample of establishment employment and payrolls is the largest monthly statistical sampling operation in social statistics." (Considering what we already know about statistics, and applying that knowledge to *social* statistics yet, this statement alone is worth reflection.) The BLS also uses data from state unemployment rolls and some smaller surveys, as well.

The rate of job growth in civilian employment during the Reagan years was somewhat sluggish in comparison with that of other presidential administrations, and from 1989 to 1991 during the Bush years, employment actually lost ground by 0.2 percent. The Reagan years produced a 2.0 percent average annual growth rate, which matches the rate of the Ford Administration but falls short of the Nixon Administration (2.2 percent) and the Johnson Administration (2.4 percent.) Clinton's hero, John F. Kennedy, presided over only a 1.5 percent average annual growth rate. A pattern is beginning to emerge.

Income growth can refer to many things. Did Clinton and Gore mean an average of workers' hourly wages? Did they mean total gains of wealth, including financial wealth and capital gains? Or did they have another measure in mind? The average annual growth in per capita disposable personal income would be a solid choice. Separate Bush and Reagan again, and Reagan fares remarkably well. Under his term in office from 1981 to 1989, this measure grew 1.8 percent. Under his predecessor Carter's term from 1977 to 1981, it grew 1.6 percent. Now let's look only at Bush. There, Clinton and Gore would have a point. Under Bush's watch from 1989 to 1991, per capita disposable personal income actually shrank .43

percent. (Regardless, from 1978 to 1990, per capita disposable personal income in the United States grew 21 percent.)

What is the pattern? What are these numbers actually saying? That Bush apparently was not good for the economy, but Reagan was? Clinton and Gore surely didn't want to say *that*, of course—they had positioned themselves *opposite* Reagan's economic policies. So they simply fused the two.

### The Tax Cut in Arkansas?

*"Clinton brags that the '91 [Arkansas] legislative session cut taxes for lower-income and middle-class taxpayers."*

> *Newsweek*
> January 20, 1992

Almost certainly, there's nothing that citizens find more infuriating about government than taxes. Perhaps we can forgive politicians their endless windy rhetoric, their trash heap of broken promises, their outward assumption of moral superiority even as they pursue wine, women, and wealthy backers, and their penchant for putting their own re-election before any other cause—but when they reach into our pockets for the funds with which to gain votes and wield power, we react as though we're being robbed.

So it isn't surprising that politicians so often offer tax cuts for certain groups of citizens. Put plainly, that's where the votes are. Although voter turnout tends to increase as income goes up, the majority of votes still come from the lower-income and lower-middle-income citizens. For this reason, politicians advertise tax cuts for them, but in order to avoid cutting spending, which also maintains large blocks of votes, they can't cut taxes on all voters. So they target their promises of tax cuts and government spending directly to the lower-income and lower-middle-income citizens in particular, gradually eliminating the payoff as they go up the income ladder and its dwindling number of voters. The votes of upper-middle-income citizens are still significant, but the votes of upper-income citizens are virtually worthless at the polls.

The money has to come from somewhere, and it comes from the small amount paid by a large number of citizens in the lower brackets and the large amount paid by a small number of citizens in the higher brackets. In this last excerpt, Clinton claimed a cut for Arkansas' lower-income and middle-income taxpayers. This makes vote-getting sense, but he didn't finish painting the picture. The total tax burden for the residents of Arkansas actually became *larger*, not smaller. The per capita state tax burden for Arkansas (not including local and federal taxes) went from $962 in 1990 to $998 in 1991. Lower-income taxes (and some middle-income taxes) went down, and spending went up.

It's a classic political ploy, but it can't be supported for long. That's because the rest of the middle-income residents and the upper-income residents saw their taxes go up, and as there simply aren't enough of them (especially in Arkansas, the fourth-poorest state in the Union in terms of personal income) to support this election tactic time and time again, taxes creep in everywhere else. In the case of Arkansas, this meant simultaneous increases in the general sales tax and in taxes on such items as gasoline and beer. In all, the total tax burden on the citizens of Arkansas increased to $272.6 million in 1991—an addition that amounted to 11.7 percent of the entire state revenues.

### Helping the Rich at the Expense of the Poor?

*"Mr. Bush proposed cutting the capital gains tax by half. Democrats have called this helping the rich at the expense of the poor."*

> The New York Times
> op/ed by William Safire
> February 27, 1992

A capital gain is money earned through the sale of an asset such as a farm, a home, a business, or stocks. As part of the 1986 Tax Reform Act, the taxation on capital gains was increased, making them subject to the same tax rate as other income. At that point, the effective capital gains tax rate jumped from 20 percent to 28 percent, an increase of 40 percent. Bush tried repeatedly to get this reduced, but as the quote above indicates, his political opponents routinely portrayed such a reduction as one that would benefit only the rich.

Let's look at the numbers and the logic to see if this is so. The spirit of the 1986 raise in capital gains taxes would appear to be that of making the rich pay more taxes. Did it work? The year before the capital gains hike went into effect, the government collected $213 billion in revenue from it. By 1991, this amount had gone down to $108 billion.

What happened? Even if the spirit of the tax hike simply had been to bring in more revenue, clearly that didn't work. It's counter-intuitive, but raising tax rates doesn't necessarily increase revenues from those taxes. In this example, when capital gains taxes are raised, citizens avoid taxes by not earning capital gains. If we don't want to give the government 28 percent of the profit from selling our home—especially when the "profit" came from simple inflation, and the house isn't actually worth any more than it ever was—we simply stay put.

At the time of the tax increase, the Congressional Budget Office (CBO) predicted that the revenue from capital gains taxes would reach $269 billion by 1991. It only reached $108 billion. This means that revenues from capital gains taxes in 1991—five years after the tax hike—amounted to only half what they were in 1985—the year before the tax hike.

At least partially because of this serious predictive error by the CBO, the national debt unexpectedly ballooned by an extra $100 billion between 1989 and 1991. Static assumptions are a common problem in predictive statistics. In this case, the CBO didn't realize that when taxes change, people change, too. Behavior that is taxed is discouraged; behavior that is not taxed is encouraged (which is the principle behind "sin" taxes, such as increasing the tax on tobacco products). This shouldn't have been a surprise. From a historical perspective, every lowering of capital gains taxes has resulted (counter-intuitively) in increased revenue from them, and every rise has resulted in decreased revenue.

The taxes themselves have economically debilitating consequences, but the error of assuming that rising rates bring rising revenues adds further weakening effects. By causing people to freeze the selling of old assets—and, other things being equal, the buying of new ones—economic dynamism is stymied, and capital is frozen into old investments instead of becoming available for new ones. That is, money stops changing hands. As an example of just one possible result in this case, between the time the 1986 capital gains tax hike went into effect and 1991, the amount of money in the United States going into venture capital for small businesses fell by two-thirds.

So does this mean that the rich managed to evade the tax man? Not at all. (Per capita, the rich pay an astonishing amount of taxes.) According to Internal Revenue Service data from 1987, 10.3 million tax returns listed capital gains. About two-thirds of them were from taxpayers with gross incomes below $50,000, and about half of this under-$50,000 group had gross incomes below $25,000.

Moreover, capital gains often represent a sudden, one-time addition to a person's net financial worth that may never be repeated. If we sell our house and retire into an apartment, yes, our tax return this year might yield a snapshot of a somewhat wealthy individual. But what did we actually gain? We may have the money, but we no longer have the house. (We have even less after we pay the tax.) And we're not going to repeat this sale year after year after year. But this same snapshot, multiplied all over the country at tax time every April, continues to provide a provocative and genuinely misleading picture of bloated plutocrats perennially at leisure while their money piles up in the bank.

Let's look at the income of taxpayers who reported capital gains, but let's subtract the capital gains itself. Data from the Joint Committee on Taxation show that of taxpayers reporting capital gains in 1985 and 1986 (before and after the capital gains tax increase) the percentage of capital gains accruing to taxpayers earning $200,000 or more fell from 45.3 percent to 25.4 percent. Similarly, the percentage of capital gains accruing to taxpayers earning $30,000 or less jumped from 12.9 percent to 26.5 percent.

Far from being the Bill Gates's of this world—and there are only a handful of them—the capital gains crowd is mainly Mom and Dad and you and I. Internal Revenue Service data from 1979 to 1983 show that only 16 percent of capital gains-claiming taxpayers claimed them every year, and 44 percent claimed them only once throughout that entire five-year period.

Another consideration is also warranted, first mentioned in the case of selling a house at a "profit" when the house isn't actually worth any more (compared to other houses) than it ever was. Capital gains are not adjusted for inflation. This means that an item (like a house) that was purchased, say, decades ago may very well be sold at a price that, when adjusted for inflation, is only equal to (or even less than) what it cost in the first place. (This is a very common situation with houses.)

Today's dollar is not quite the same as yesterday's, has less relationship to the dollar of three years ago, and is only remotely akin to the dollar of three decades ago. What gives a dollar value is our ability to exchange it for goods or services, and the amount of goods or services that we can buy with a dollar shrinks steadily with inflation. So you may have bought a house for $20,000 that inflated in value to $70,000 over the years, but so did the price of bread and milk and everything else. When you sell it to retire to an apartment, you gain nothing at all, but you'll have to pay capital gains tax on the $50,000 "profit" anyway. Such is the world of the "wealthy."

Economists adjust for inflation by multiplying the dollar of any given year in the past by the percentage rise in the Consumer Price Index (CPI), but the technique is far from comprehensive. The CPI is calculated by choosing a set "market basket" of items whose changes in price are measured over time; these changes are then assumed to represent the change in the purchasing power of the consumer's entire budget. Calculating inflation is an inherently loose procedure. One problem with the CPI is the assumption that any one basket of goods could reasonably represent a useful measure of the standard of living of so vastly varying a populace as 250 million people. (Fax machines affect some of us very significantly; they don't affect others at all.)

Another problem is that there is no numerical way (for this very numerical of concepts) to measure changes in the quality of goods, which is an essential key to our standard of living. (Consider the difference between the icebox of yesterday and the refrigerator of today.) A third problem is that there is no way to account for changes in desire, which is not the frivolous consideration that it may seem at first. (Early in this century, our great-grandparents had little desire for electricity; today they desire it very much.)

And what about the concept of need? Does any human animal really need fax machines, modern refrigerators, or even electricity? Obviously not. But we'd surely call anyone who lacked electricity in his home "needy." When my mother grew up, her mother cooked everything from scratch and made her own clothing. Her family also felt blessed that they had everything they needed, and everyone was content. The only luxury afforded the children was a chocolate bar once a week, and that was greatly appreciated. But imagine the reception we would get from the poorest segment of our population if we were to cut welfare back to a small fraction of its current budget and instead truck people an abundance of free supplies like milk, flour, sugar, and bolts of cloth.

Another problem with the CPI is that inflation occurs in different parts of the market at different times; prices don't join forces and march up together the way the CPI implies. Many necessities (such as water) remain incredibly cheap while many luxuries (such as soda) have soared by comparison. Also, relative prices in high-tech areas of the economy are highly misleading over time. For example, this year's top-performance computer costs a small fortune. But if you buy last year's top-performance computer, it costs a small fraction of that. And while last year's model may cost the same as a top-performance computer of five years ago, the newer model may perform ten times as many functions ten times as fast. The BLS does make an effort to account for this sort of component with estimates that it calls "hedonic quality adjustments" (which we might define as the "pleasure factor"), but such attempts are understandably weak.

Occasionally, the BLS recognizes such great flaws in the CPI that they actually make a change in the method of measurement. (Although an appropriate step, this change then creates a whole new source of confusion and error.) One such change, in 1983, makes any previous poverty or inflation measure suspect, including new measurements with old numbers.

Before 1983, in order to reflect month-to-month changes in the price of housing, the CPI reflected month-to-month changes in the special one-time costs incurred when buying a house—even though no one buys a new house every month. By switching to reflect month-to-month changes in the cost of rental housing instead (for the same household budget), the amount of inflation turned out to be lower than it was thought before. Thus, any comparative cost-of-living analysis that uses (unadjusted) pre-1983 CPI data overstates inflation.

In 1994, the BLS again considered changing its method of calculating the CPI, an additional indication that the figure can't yet be relied upon to give a true picture and indeed may never do so, considering its inherent weaknesses. This is of major importance because flaws in the CPI translate into flaws in nearly every aggregation of economic variables for comparison over time; an accurate adjustment for inflation is imperative for such analysis.

Specifically, the BLS suspects that the CPI may overstate inflation by as much as 0.6 percentage points every year. Part of the problem, the Bureau acknowledges, is with the conceptual difficulty of assuming that a market basket set back in the years 1982 to 1984 adequately reflects the buying patterns of people a decade later. (Note that buying patterns have little to do with acquiring the basic necessities of life. In this country, those necessities are both inexpensive and plentiful.)

Another part of the problem is a mathematical misconstruction that dramatically affects inflation calculations when a price fluctuates up and down. For example, a rise in the price of a commodity from \$1 to \$1.25 registers as a 25 percent increase because the 25¢ rise is 25 percent of \$1.

But if the $1.25 price then drops back down to $1, it registers as only a 20 percent decrease because the 25¢ drop is only 20 percent of $1.25. The 5 percent increase (the 25 percent increase minus the 20 percent decrease) stays in the CPI even though the price is exactly the same as it was before.

But regardless of the obvious flaws of the CPI, capital gains taxation that is not adjusted for inflation is even more flawed. Not only do you find yourself taxed on a "gain" that doesn't exist, you may even find yourself taxed on a loss. (And plenty of us do; it's a common scenario.) Suppose you bought a house for $20,000 that inflated in value to $70,000 over the years, but you're only able to sell it for $60,000. You have a $10,000 loss, but you'll have to pay capital gains tax on a $40,000 profit, instead. The world of the wealthy can be a hostile one, indeed.

### The Top One Percent Got 60 Percent of the Growth?

*"Bill Clinton . . . has been repeating a startling statistic: Those whose incomes put them in the top 1 percent got 60 percent of the economic growth in the past dozen years."*

*The Wall Street Journal*
April 8, 1992

This quote exemplifies one of the most vivid images that Clinton successfully created to hold against Bush—that the combined years of Bush and Reagan defined a time when the super-rich became super-richer far out of proportion to their number, presumably at the expense of everyone else.

First, it again should be noted that throughout the dozen years in question, it was a Congress controlled by Democrats that played the key role in economic policies, particularly in taxing and spending policies, and that Bush and Reagan followed different courses from each other.

But even leaving those points aside, Clinton's statement is still misleading, despite the fact that it contains two specific numbers. For one thing, it personifies the top 1 percent, as if the same people—the same names—steadily made economic gains over and over again during the years from 1980 to 1992. This isn't so, and the error is rooted in a fallacy that is a common element in numerical misunderstanding, a fallacy that we encounter repeatedly as we examine the economic statements and predictions made by politicians: the fallacy of the static economy.

Labeling a top 1 percent or 5 percent or 20 percent over time does not identify the same individuals over time, and the smaller the percent, the larger the difference. (And the larger the percent, the smaller the difference. That is, if we label a group of 100 percent over time, we will indeed identify most of the same individuals, although plenty are born, move, and die. But if we label a group of only 1 percent of those individuals, there are many more people available who might occupy one of the top-1-percent positions at some point in time, however briefly.)

So the 2.5 million people who composed this top 1 percent in 1992 were different from the people who composed it in 1991, and the 1991 people were different from the people who composed it in 1990, and the 1990 people were different from the people who composed it in 1989, and so on. After all, times change, and so do people. Folks get older as the years go by, they work harder, they gain experience, they rise into the ranks of the successful, they retire, and they die—not to mention the effects of good luck and bad luck along the way. And in the United States in particular, the top-income groups are more fluid than in any other country in the world. The best studies of income mobility in this country are done by "quintile analysis," that is, the division of American households into fractions of fifths, based on income. (This comes from the Latin word *quintus*, or "fifth.") But quintile analysis is fraught with misunderstanding, error, and fallacy.

In particular, it must be remembered that the division is in households, not individuals. When we see quintile analyses, we shouldn't assume that any one quintile actually contains 20 percent of the individuals. For example, in 1989, the "bottom quintile" of income contained only 14 percent of the individuals in the country, while the "top quintile" contained 25 percent.

This is because households are bigger in the highest-income quintile. Although there are a significant number of single-parent families with a large number of children in the lowest-income quintile, this is more than offset by the fact that the lowest-income quintile also contains more single people living alone and elderly people who have lost their partners and don't live with their children, either. In that particular year, the average size of a household in the highest-income quintile was 72 percent larger than the average size of a household in the lowest-income quintile.

There are many other numerical complications involved in the use of quintile analysis. For example, an increase in the income of a household in any of the lower four income quintiles may increase averages in other quintiles more than its own quintile. Here's an illustration for ten households:

---

Number 1  Income Quintile

    One household earning $100,000 a year

    One household earning   90,000 a year

Number 2  Income Quintile

    One household earning   80,000 a year

    One household earning   70,000 a year

Number 3  Income Quintile

    One household earning   60,000 a year

    One household earning   50,000 a year

Number 4  Income Quintile

    One household earning   40,000 a year

    One household earning   30,000 a year

Number 5  Income Quintile

    One household earning   20,000 a year

    One household earning   10,000 a year

---

Let's say the aspiring writer earning $10,000 a year as a part-time waiter finally sells his book for an advance royalty of $100,000. As the number 5 Income Quintile now averages $15,000 a year ($20,000 + $10,000 = $30,000 ÷ 2 = $15,000), that's a staggering increase to a $65,000 average, right? ($20,000 + $110,000 = $130,000 ÷ 2 = $65,000). No. Instead, the writer/waiter moves to the number 1 Income Quintile, rearranging them as follows.

---

Number 1  Income Quintile

    One household earning $100,000 a year

    One household earning  100,000 a year

Number 2  Income Quintile

    One household earning   90,000 a year

    One household earning   80,000 a year

---

Number 3  Income Quintile

    One household earning    70,000 a year

    One household earning    60,000 a year

Number 4  Income Quintile

    One household earning    50,000 a year

    One household earning    40,000 a year

Number 5  Income Quinttile

    One household earning    30,000 a year

    One household earning    20,000 a year

So the highest-income quintile (and all the others, for that matter) can be said to have benefited at the expense of the lowest-income quintile. Of course, that warps the situation into a gross misunderstanding of fact, but it's done all the time.

What happens when the highest-income quintile takes the same sort of jump? Let's go back to the original quintile illustration. This time we'll say that the aspiring artist earning $100,000 a year as a physician finally sells one of her paintings to a grateful patient for the sum of $100,000. What happens to all the other quintiles? Here's the rearrangement:

Number 1  Income Quintile

    One household earning $200,000 a year

    One household earning    90,000 a year

Number 2  Income Quinttile

    One household earning    80,000 a year

    One household earning    70,000 a year

Number 3  Income Quintile

    One household earning    60,000 a year

    One household earning    50,000 a year

Number 4  Income Quintile

    One household earning    40,000 a year

    One household earning    30,000 a year

Number 5  Income Quintile

One household earning    20,000 a year

One household earning    10,000 a year

---

The number 1 Income Quintile was the only one affected. No matter what happens, gains seem to pile up at the rich end of the scale. In this case, it's because any increase in the highest-income quintile increases the average income for that quintile alone, because there's no higher one to which the household can move.

Regardless, we must deal with the numbers we have, and the Census Bureau has collected numbers on movement between quintiles. Further dispelling the notion of the highest-income quintile comprising the same people—the same names—from year to year, the Bureau reported that between 1985 and 1986, 23.7 percent of the households in the highest-income quintile moved down into a lower one. That was not a statistical aberration; 24.3 percent also moved down between 1987 and 1988.

Two examples don't make a trend, but these numbers still reflect what most of us already know, even if we'd rather not think about it: Yearly incomes don't provide lifetime security. Success waxes and wanes, and even large financial gains are often confined to a limited period. When Clinton spoke of "those whose incomes put them in the top 1 percent . . . in the past dozen years," he was actually referring to far more people than this statement implies.

The rest of his statement (". . . got 60 percent of the economic growth in the past dozen years") is even more faulty. The reported source of Clinton's information was an analysis of the growth in average family income, written by economist Paul Krugman of the Massachusetts Institute of Technology and published in *The New York Times*. In the first place, Clinton's statement was made in 1992, but the study covered the years 1977 to 1989, so it comprised the Carter and Reagan presidencies, not the Reagan and Bush presidencies. Second, the study was based on average family income, not individual income, but Krugman failed to adjust for the fact that average family size declined by 10 percent over that time period.

Third, considering a gain in average family income accrued to given quintiles can be misleading. This is the way economist Michael Boskin puts it:

Mr. Krugman's calculation bears no relation to how gains from economic growth are distributed. Suppose an economy has two workers with annual incomes of $20,000 and $30,000 respectively. A few years later, these same workers earn $30,000 and $40,000 respectively, and two new workers obtain jobs earning $20,000 each, all figures adjusted for inflation. Total real income increased by $60,000. One-sixth accrued to the worker now representing the top 25 percent of income earners. One-sixth accrued to the worker in the next 25 percent of income earners. Two-thirds accrued to the bottom 50 percent of earners. Average income increased by $2,500 (from $25,000 to $27,500) while average income in the top half of the distribution increased by $5,000 (from $30,000 to $35,000).

[But] Mr. Krugman's calculations of the gain in average income would indicate that people in the top half accounted for 100 percent of the gains from economic growth—of the gain in average income—even though the two original workers received equal raises, one moved from the bottom half to the top half of the distribution, and two new workers found jobs and now constitute the bottom half.

Here's how that looks:

|  | Before change: | After change: |
|---|---|---|
| Upper Half | $ 30,000 | $ 40,000 |
|  |  | 30,000 |
| Lower Half | 20,000 | 20,000 |
|  |  | 20,000 |
| Total Income | $ 50,000 | $110,000 |

In other words, before the change, the bottom half averaged $20,000, and after this clearly beneficial change to every worker involved, the bottom half still averaged $20,000.

The fact is that the Joint Economic Committee of the U.S. Congress has gathered numbers from Census Bureau data of real average family income (in constant inflation-adjusted 1991 dollars), and these numbers show that when the presidential economic policies that Clinton was running against were in effect (from 1982 to 1991, the nearest year for which data was available), the highest income *quintile* (composed of a changing 20 percent of households, not a static 1 percent of individuals) accounted for 63.6 percent of the total income gain. But because we've already seen how those gains can misleadingly appear to pile up at the top, even this fact means very little.

### Save Social Security with a Cutoff in Benefits?

*"Perot says he could save $20 billion a year by ending benefits for 'folks that don't need it,' including himself."*

> *The New York Times*
> June 15, 1992

What generosity. But most of the other "folks that don't need it" aren't quite billionaires. The Bush Administration replied to Perot with calculations by the Actuary of the Social Security Administration which showed that net savings of $20 billion to the Social Security program would not be accomplished by withholding payments above a specified income level until that cutoff level went as far down as a household income of $56,000 a year. (The figure is net because Social Security payments are taxable above a certain level. If payments are cut off, tax revenue is lost.) Bush's people apparently thought that this number stood as an unanswerable challenge to Perot, but it doesn't address the real issue, which is one of the most serious issues this country has ever faced. According to the numbers, the Social Security program is headed for a fall.

Most of us think of Social Security as something very different from a welfare program. We believe that we pay in our share, and that someday we will be entitled to take that share back out—a sort of forced saving for retirement, generously supplemented by our employers, who are required to contribute an equal share on our behalf. If that's the case, there's no real reason for Perot to feel guilty about receiving money that he doesn't need. After all, it's just his own money (and his employer's) returning to him, right?

Wrong. Social Security is a tax that helps to fund a social-welfare program for broad groups of people (not just older people), and although the
formula for contributions is related to income, and the formula for payments is related to contributions, it is not a retirement program in which
you someday receive back only what you (and your employer) contributed,
plus interest. If that were the case, it wouldn't be failing. In fact, Social
Security pays out vastly more than it takes in and has spread its wings far
beyond its original intent. People who currently receive Social Security
payments are actually receiving money directly from today's workers.

Politicians like to talk about the Social Security "trust fund" that we've
accrued because of increases in Social Security taxation during the 1980s,
but the truth is that this "trust fund" is spent instantly on other government programs. The money is lent to the federal Treasury and replaced
with federal Treasury bonds—in other words, government I.O.U.'s.
When the Social Security system is desperate for cash because current tax
intake doesn't cover current payout, taxes are raised on the current working population.

In an attempt to assess the program's future needs, the Social Security
Administration (SSA) makes three different actuarial projections based on
"optimistic," "intermediate," and "pessimistic" assumptions. By 2030,
according to "intermediate" assumptions about future birth rates, mortality rates, and income growth, the federal government will be paying out
between $2 trillion and $4.8 trillion more in Social Security and Medicare
(Part A) benefits than it will be taking in through taxes. This would require
total Social Security taxes to rise to 25.4 percent of the payroll for the
entire working population just to get through that one year.

So you will not get back your own (and your employer's) money.
Rather, you will get only what future taxpayers can afford to pay you. In
terms of actual cash on hand at present, the "trust fund" is only able to
make Social Security payments for another thirteen and a half months.
*Your* money has long gone.

However, intermediate assumptions are not necessarily the most realistic ones simply because they are midway between the two extremes. In
real life, economically speaking, the pessimistic projections often are the
more realistic ones. Two examples in recent experience: Increases in real
wages have been lower than the Social Security Administration's most
pessimistic projection, and annual increases in the Consumer Price Index
have been higher than the most pessimistic projection.

Moreover, maintaining the current "surplus" in Social Security is not without great current cost. An analysis by the Institute for Research on the Economics of Taxation shows that the Social Security payroll tax hikes of 1988 and 1990 have increased the tax burden on working Americans by $500 billion. And if current birth and mortality projections stay steady, by the year 2030, there will be only two workers paying taxes to support every individual receiving Social Security payments.

Even worse, if the government were actually obligated to pay back every taxpayer's contribution to Social Security, consider this: For just the retirement portion (not including Medicare and hospital insurance) as of the beginning of 1990, this accrued liability—the amount of money needed to pay off, with interest, all current liabilities for everyone in the program—was $6.7 *trillion.* The "trust fund" at that point had assets worth $189 billion. That's a big gap. Within our lifetimes, something's undoubtedly got to give.

### The Declining Job Base?

*"In a June 5 interview with the* Los Angeles Times, *. . . Perot cites, as he often does, the 'declining job base' as the nation's major economic problem."*
Newsweek
June 22, 1992

Whatever could our tough-talking Texan have been complaining about here? He couldn't possibly have meant that the number of jobs existing in America was truly declining. From 1982 to 1989, before the beginning of the last recession, official figures state that more than 4.5 million new businesses were created in the United States, supplying new jobs to about nineteen million Americans. New businesses and, indeed, entire new industries are continually being created in a vibrant economy. But Perot has built a reputation as being a defender of older, more labor-intensive industries, such as the kind that he claimed the North American Free Trade Agreement (NAFTA) might threaten. In this role, he maintained that many jobs are being lost as these older industries are farmed out to foreign lands where labor is cheaper.

Never mind that Perot's statement wasn't credible if it referred to an actual countable loss in available jobs in a current time period versus any given earlier period, and that without naming a time period, it can't be verified. He did give attention to a subject that had (and still has) great emotional resonance for many Americans, as witness Perot's success in becoming a significant part of the public (and very political) debate over the economic wisdom of NAFTA.

If references to a declining job base are to make any sense at all, they must refer to losses of jobs in certain given industries. But to interpret this as a danger to the economy as a whole requires looking at only one side of the job ledger. Of course, jobs disappear; but new ones are being created at the same time. To worry about the number of jobs that no longer exist is to miss the key to economic advance: being able to be more productive with fewer resources. That generally means less human labor as well. To lose jobs—to have a "declining job base"—because new technologies allow the same or greater production to be effected with fewer human workers—is an economic *benefit,* not a drawback. Those resources, human and otherwise, once needed to produce any one item and/or meet any one need are thereby freed to produce other items and/or meet further needs.

For example, if we still used the outmoded switchboard technology of the early days of the telephone, all the women in the country between the ages of eighteen and sixty-five could probably be gainfully employed as Ernestine-style switchboard operators to handle the immense volume of calls currently produced in America. Should we lament the declining job base in switchboard operators created by the new telecommunications technology? Or should we instead cheer that we can still make our phone calls and have all those people employed doing other things at the same time?

Putting a stop to a declining job base can easily be done merely by cutting back in the use of labor-saving technologies. We could employ all able-bodied men in carrying fruit across the land simply by mandating an end to the use of trains and trucks for that purpose. Or we could all be meaningfully employed as yeoman farmers merely by ending the use of modern agricultural technology. (Fidel Castro has more details on this methodology.)

Yes, we'd all have jobs growing food; but then, food would be *all* we'd have. As an example of the nature of job loss, consider that in 1910, there were 13,555,000 people working on farms; by 1970, this number had fallen to 4,523,000. Should we lament the declining job base in farming?

No, because the number of people accommodated per farm worker had risen from seven to seventy-nine over that same period, thus freeing all those workers to produce other things—like sophisticated telecommunications technology. And that's the real purpose of jobs in an advancing society—not just to keep people busy.

## There's Something Wrong With Our Tax Code?

*"[T]here's something wrong with our tax code, if your income went up 65 percent in the 1980s and your taxes went down 15 percent."*

> The New York Times
> quoting Bill Clinton's speech
> June 23, 1992

Why? Would Clinton think something is wrong with our health care if our health became much better, but our medical expenses went down? That is, why is it inherent in the concept of betterment that we must pay more for it? Even more seriously, why is it inherent in the concept of *self*-betterment that we must pay more for it? In this case, a puritanical moral attitude is presented as mathematics. Here is the point: Clinton artificially juxtaposes these up/down percentages to make us believe that if they don't go up together and down together, then "there's something wrong." But the numbers don't justify that conclusion. It makes no sense. We should all be delighted if everyone's income went up 65 percent and their taxes went down 15 percent.

It's a shock to people when they first begin to see their hard work pay off and then personally discover the weight of the tax burden that they so willingly wished on the "rich" when they were younger. There are two basic ways in which a tax burden can grow as income increases, and our government imposes the harsher of them, scaling back the rewards of any American who follows society's advice to work hard and improve his or her lot.

One of the ways is for all citizens to pay a set percentage of their income in taxes. This is usually called a "flat tax." With a flat tax rate, as income increases, so do taxes. Say that rate is 15 percent. Putting aside all other considerations (like exemptions and deductions) the person who earns $15,000 would pay $2,250 in taxes, the one who earns $35,000 would pay $5,250, and the one who earns $55,000 would pay $8,250.

Our country does it the other way, increasing the *percentage* as income grows. This is usually called a "progressive tax." With a progressive tax rate, as income increases, taxes balloon. Say that the first rate is 15 percent, the second rate is 28 percent, and the third rate is 31 percent. All other factors aside (including marginal rates, which we'll address shortly), the person who earns $15,000 would pay $2,250 in taxes, the one who earns $35,000 would pay $9,800, and the one who earns $55,000 would pay $17,050. The difference between a flat rate and a progressive rate is stunning to the taxpayer new to this.

These were the actual rates in 1995, plus a fourth and fifth tax rate (36 percent and 39.6 percent), but they were *marginal*, as are all tax rates in this country. That is, throughout the first tax bracket, all taxpayers pay 15 percent; in the second bracket, additional income (beyond the first bracket) is taxed at 28 percent; in the third bracket, additional income (beyond the second bracket) is taxed at 31 percent; in the fourth bracket, additional income (beyond the third bracket) is taxed at 36 percent; and in the fifth bracket, all additional income (everything beyond the fourth bracket) is taxed at 39.6 percent.

So when you consider the progressive tax system currently in this country, Clinton's seemingly clear statement suddenly becomes cloudy. Let's look at the numbers more closely. Does he mean that the total *dollar amount* of taxes paid by this hypothetical American success story was 15 percent lower than it would have been without the changes in the tax code during the 1980s? Or does he mean that her *tax rate* was 15 percent lower? Those are two very different things.

Let's say our physician was earning $108,301 in 1980—and paying taxes at the actual then-current 70 percent marginal rate—and a decade later, was earning $178,697 a year, a 65 percent increase. If her "taxes went down 15 percent" over the same period, what exactly does this mean?

If her "taxes went down 15 percent," it might mean that her total *dollar amount* of taxes in 1980—that is, $55,698 in taxes on an income of $108,301—decreased by 15 percent by 1990—that is, went down to $47,343 in taxes on an income of $178,697. But what if she were paying a ridiculous amount of taxes in 1980? (We're only considering federal income tax here; state and local income taxes are not included.) Wouldn't a lowering of those taxes make good sense? Regardless, that 15 percent decrease simply didn't occur. Her actual tax bill in 1990 was closer to $54,089, less than a 3 percent decrease.

Then again, if her "taxes went down 15 percent," it might mean that her *tax rate* went down, instead. In that case, because she owed $55,698 in taxes on an income of $108,301 in 1980 (a true average rate of 51 percent), she then would have owed $77,465 in taxes on an income of $178,697 in 1990 (a true average rate of 43.35 percent, which is 85 percent of the previous rate of 51 percent). So her dollar amount of taxes actually would have gone *up* (from $55,698 to $77,465) by 39 percent.

Even worse, this analysis discounts the very important factor of inflation. When an income grows by any percent during times of inflation, it does not grow by that same percent in real purchasing power. It always grows less—sometimes much less. (Over a decade, income growth can look misleadingly impressive on paper; in the case of our physician, the growth was only 5 percent per year.) Real purchasing power may even shrink. But as if that weren't bad enough, while this is going on, inflation silently carries us into a higher tax bracket.

Here's an example: Let's focus on three taxpayers in the pre-Reagan year of 1980. The first one earned a taxable income of $18,200 and paid $3,565 in taxes; his effective tax rate is 20 percent. ($3,565 ÷ $18,200 = 20 percent). The second one, in the next bracket, earned $23,500 and paid $5,367; his rate is 23 percent. The third one, in the next bracket, earned $28,800 and paid $7,434; his rate is 26 percent. But now let's say that Carter had won the election instead of Reagan, and he kept taxes at the same rate (instead of lowering them, the way Reagan did.) An inflationary four years passes, and the taxpayers receive raises from 5 percent to 7 percent each year. Now the first taxpayer is earning $23,500, the second one is earning $28,800, and the taxpayers in higher brackets move up likewise. But inflation has wiped out the gain, and real purchasing power remains about the same as it was in 1980. Are these people stable in purchasing power?

No. They've quietly floated into higher tax brackets. The first taxpayer no longer pays 20 percent of his income in taxes; he now pays 23 percent. The second taxpayer no longer pays 23 percent of his income in taxes; he now pays 26 percent. (And so on with the third taxpayer and those in higher brackets.)

But putting aside the question whether flat taxes or progressive taxes make better ethical sense, we must look at the results of lowering tax rates in general to determine whether Clinton's statement made good *economic* sense. And interestingly, when we do that, we see that if higher-income people paying a larger share of the American tax burden makes good ethical sense (whatever the method), the 1980s look far more productive to that end than Clinton's statement implied.

In fact, the share of the total federal income-tax burden paid by the wealthiest 5 percent of Americans increased by 17.6 percent during the 1980's. This share rose from 36.8 percent to 44.1 percent, even though effective federal income tax rates for them did indeed go down. (For example, the rates for the wealthiest 20 percent of Americans went down 8.8 percent.) This was possible partly because the effective federal income-tax rates for everyone else were going down even faster.

Moreover, looking superficially at changes in tax policy—for example, a lower tax rate for the wealthy—ignores the changes in human behavior that this tax policy brings about. Changes in tax rates affect human behavior before they affect tax dollar results. When high marginal tax rates were cut and loopholes were eliminated—both of which occurred during the 1980s—the wealthy not only earned more income, they also earned more taxable income. For example, from 1981 to 1988, tax payments by the top 1 percent of taxpayers increased by 51 percent.

So, as a result of tax-policy changes in the 1980s that Clinton assailed, wealthier Americans paid both a larger share of the federal tax burden and, according to economist Lawrence Lindsey, the government collected $11.5 billion more from taxpayers earning an annual income of $200,000 and up than it would have collected had those changes not occurred.

Didn't Clinton know the financial reality? Or was he just using jealousy to get votes? "Redistribution of wealth" has an appeal to many voters because they assume they'll gain from it. But it doesn't work that way in real life. There isn't enough wealth to make a difference after it's divided by 250 million people.

## Put Everything on Hold Until the Deficit is Slain?

*"Shouldn't everything else be put on hold until the deficit monster has been slain? The budget deficit, estimated at $372 billion for fiscal 1992, is serious, but it is not our only—even our main—economic problem. . . . If we balanced the budget by 1996, instead of running the deficit the Congressional Budget Office projects, a reasonable estimate is that our long-term growth rate would rise by only about 0.25 percent."*

The New York Times
op/ed by Paul Krugman
August 4, 1992

Although they love to spend money, deficits are the *bête noire* of every politician. All three candidates in 1992 felt it necessary to demonstrate fiscal probity by coming up with a deficit-reduction plan to show that they could bravely slay the growing beast at least quickly, if not without pain. On the surface, this makes good sense. It is an obvious tenet of good business management that outgo cannot exceed income—at least not for long—and we should expect no less a fiscal standard from the executive manager of the federal government than we do from the executive manager of a business. But not spending more than earnings on an arbitrary year-to-year basis isn't always the way businesses operate, and few of those televised federal deficit number quips are as clear-cut as they seem at first.

Mr. Krugman's assurance about the increase in economic growth rate that we could expect from cutting deficit spending should be taken, like most economic predictions, with a mountain of salt. Economists' abilities to foresee the results of given changes in the economic environment are legendarily weak, and for good reason. Local wags say that if you took all the economists in Washington and laid them end to end, they'd still all be pointing in different directions.

This joke is rooted in fact, but the facts have little to do with economists personally. Even if economics is a science, the predictions of its "scientists" only rarely come true, at least partly because they lack an experimental method like that of most other sciences. In real experimental science, the investigator must have one—and only one—independent variable (in Krugman's example, deficit spending) to ensure that the change in the dependent variable (in Krugman's example, the economic growth rate) can indeed be attributed to the behavior of the independent variable. In other words, all other variables must be controlled and kept constant. But in considering the economy, these "other variables" are actually all the choices and resultant actions of every member of society. While economists strive to emulate the numerical precision of the physical sciences wherever possible, the essential difference in the nature of their subjects ultimately renders this goal unattainable.

Because the study of economics involves so many elements that can be counted—such as deficit spending, interest rates, monetary exchange rates, and changes in employment—many citizens are misled into thinking of economics only as a science of numbers. But in action, economics is a social science, not a physical one. The alert citizen would do well to re-member this the next time he or she hears a politician or pundit make a numerically precise assertion about what effect any change in one economic variable will have on some other economic variable.

Economic growth depends on countless decisions about working, saving, spending, investing, and so on, made by some 250 million individuals, none of whom are inanimate particles whose reactions to any given stimulus can be known beforehand. To make an assertion about how a change in deficit spending would necessarily affect future economic growth is hubris indeed, and hubris unsupported by the general success rate of this sort of economic prediction. This is especially pertinent when we are uncertain about the number to begin with—in this case, the federal budget deficit. One would presume that this figure is a clear-cut accounting of expenses minus revenues in a certain fiscal year. But one would presume wrongly. There are, in fact, major complications.

One complication is that current Social Security fund figures are being used to offset deficit figures, even though this money rightly should be considered sequestered for the payment of Social Security obligations. Counting it as money available for other spending during the current year violates the principle of Social Security as a trust fund. If this practice had been changed for the year 1992, $49 billion would have been added to the official deficit figure.

Another complication that makes federal budget deficit figures so different from the profit-and-loss statements issued by corporations is that the government doesn't attempt to separate consumption expenses from capital expenses. This is a controversial point. There's no argument that a great deal of pork will be found in any government investment. Some even maintain that nothing the government does should be considered an investment; they advocate that any government spending actually should be subtracted from Gross Domestic Product figures because all government money is, in reality, either taken away from productive taxpayers or borrowed out of a finite fund of potentially productive capital.

But a portion of government spending, such as that earmarked for infrastructure and research and development, is at least intended in the spirit of capital expenditures (whether it's a porky project or not), and, after all, not all private investments have a productive payoff, either. Of course, this is not a black-and-white issue, but deficit figures are treated as though *they* are. At any rate, if such an adjustment is made, in keeping with the accounting practices of most corporations, $248 billion would have been subtracted from the official 1992 deficit figures.

Of course, none of this changes the fact that the government is spending more than it is collecting in taxes, but is it necessarily better for government to tax for every dollar it spends each year? Contrary to our instincts, this is a debatable issue. There are worse things for an economy than having the government spend more than it takes in. For example, taxing for the full amount of federal spending could have far more dire effects than paying for that spending with a combination of taxes and deficit spending, because the effect of higher tax rates is to discourage greater work and productivity in the immediate term, while the immediate ill effects of deficits aren't always so severe.

According to orthodox economists, those effects are supposed to include higher interest rates as the government increases demand for borrowed funds in a monetary supply that is assumed to remain about equal. Yet inflation-adjusted interest rates fell through the mid-1980s even though deficits rose. There's always the possibility interest rates might have fallen even more if not for those deficits, but this all falls into the larger lesson of the inherent unpredictability of the economy because of the plethora of influences at work. There were clearly other factors ameliorating the effect of the federal government crowding of funds.

And as an example of how little politicians can be trusted with their predictions about matters such as the deficit, it can be noted that even Mr. Krugman's learned guesstimate about the projected federal deficit for that year was off by $27 billion.

### Raising Taxes on Only the Rich Is a Fraudulent Claim?

*"Clinton's claim that he will only raise taxes on the rich is a fraud. There are not enough incomes above $1 million to fund the Democratic spending machine. . . ."*

> The Wall Street Journal
> Memo from Jack Kemp, Vin Weber,
> Newt Gingrich, Connie Mack, Trent
> Lott, Bob Kasten, Malcolm Wallop
> August 11, 1992

This is an example of how putting statements near each other makes them appear to be related, when in fact, they are not. Yes, Clinton stated that he would only raise taxes on "the rich." But no, he didn't state that "the rich" had a numerical lower boundary of a $1 million income. In fact, in the first presidential debate, Clinton specifically stated, "The tax increase I have proposed triggers in at family incomes of $200,000." Other analysts had said that to get the money Clinton needed, his tax increases would have to affect families with far lower incomes. But Clinton's statement struck a chord with many voters, who may have seen $200,000 incomes as belonging to "millionaires" rather than a family, probably with at least two working members.

If we define "the rich" as millionaires, it is indeed impossible to run the United States federal government by forcing them to pay for it at *any* percentage of their incomes. In fact, a 100 percent tax (which would have to be a one-shot deal—no one is going to continue working and earning money only to see every cent of it confiscated by the government) on the earnings of America's millionaires would be enough to run the government for a mere thirty-three days.

Aiming our sights a little lower, down closer to the bracket Clinton was targeting, if we add a 100 percent tax for all the nation's *half*-millionaires, it would be enough to run the government for only fifty-five days. And confiscating on down to the level that Clinton himself defined as "the rich" (that is, families earning at least $200,000 yearly), we'd still be in trouble. A 100 percent tax on *them* would be enough to run the government for only 108 days—after which time it would all come crashing down.

In fact, in order to balance current levels of federal spending for just one year, the government would have to seize 100 percent of the earnings of every family with a combined income of over $180,000 per year.

American millionaires aren't a rich source for taxing purposes—they earn less than 4 percent of all taxable income. But because they're rich, they become targets. The problem with forcibly appropriating their funds is that there just aren't enough of them to make a difference. In 1991, there were only 52,000 millionaires in the U.S., which is only 0.05 percent of all taxpayers.

### The Deficit Will be Cut in Half?

*"Clinton says he can save $26 billion in wasteful spending and that the budget will increase by no more than $64 billion by 1996. He says the deficit will be cut in half over the next four years and continue to decline thereafter."*

Newsweek
September 7, 1992

The list of spending cuts in Clinton and Gore's book does indeed add up to around $26 billion ($26.09 billion, to be precise, which is always amusing at this level of dollars). The two candidates break this down into categories such as "Intelligence cuts," "Administrative savings," "Cut White House staff by 25 percent," "Reduce overhead on federally sponsored university research," "Reduce special-purpose HUD [Housing and Urban Development] grants," and "End taxpayer subsidies for honey producers."

But in political language, a "spending cut" isn't always a spending cut. Instead, it may be a reduction in the rate of growth. In fact, in Clinton's first five-year budget projection, the only areas that were actually cut (the way most Americans define this word) were defense and interest. All other areas of domestic spending would rise $10 billion over a four-year period, and this rise was on top of the $245 billion increase already in the previous year's budget. But even so, supposed spending cuts were only one side of Clinton's deficit-reduction ledger. Tax hikes were also featured. In 1993, Clinton vowed to gain $17.8 billion by a surtax on "millionaires" and increasing the income-tax rates on the top 2 percent of American income-earners.

Clinton follows the lead of the Congressional Budget Office (CBO) in his method of calculating how much revenue any given tax increase will produce. But the CBO method has a glaring flaw that goes straight to the heart of the reason government predictions are so often grievously mistaken. The CBO indulges in what is known as "static analysis"—the assumption that any set rise in tax rates will generate a like rise in revenues, as if people's behavior and earnings after the tax hike will remain precisely the same. This is disingenuous. The history of capital gains tax hikes and luxury taxes are good examples of the shortcomings of static analysis.

The 1990 budget agreement included a "luxury tax" on such items as boats and airplanes, and the congressional Joint Committee on Taxation estimated a five-year gain of $1.5 billion. It made sense to them at the time. After all, only rich people purchase such items, so only the rich would be affected, and who cares, anyway? They've got some money, they don't have that many votes, and the taxes would bring in a nice extra income stream.

In reality, the tax cost 9,400 blue-collar jobs in those industries. After all, it is not "the rich" who *make* boats and planes. But because the government didn't realize that upper-income Americans are already heavily burdened with taxes and might purchase fewer luxury items if their costs rose, jobs were lost, income tax from those jobs was lost, and the net effect for the first year was the loss of an estimated $14.1 million. So Clinton's claim to cut the deficit in half, based on static analysis, would almost certainly prove to be wrong because the tax hikes would not bring in the expected revenue.

### Arkansas Led the Nation?

*"Bill Clinton hasn't called it the Arkansas Miracle, but his Razorback economy sits near the top. . . . Despite the painful U.S. recession, Clinton's Arkansas has led the nation in job growth and home-price appreciations over the past year. And for the 12 months ending in March, Arkansas' real per capita income grew 1.7 percent, the eighth best performance among all the states."*

U.S. News and World Report
September 14, 1992

Here, the number "12" is the most significant. Why? Because Clinton was governor of Arkansas for twelve of the fourteen years *prior* to that period, so what happened during that one-year period is not the most relevant consideration. Other numbers that describe the Clinton eras— he was the governor from 1979 to 1981 and again from 1983 to 1992— will help illuminate the whole picture.

From 1980 to 1990, a period in which the national percentage of Americans living below the poverty line fell from 13 percent to 12.8 percent, the percentage in Arkansas rose from 14.9 percent to 17 percent. In addition, the number of welfare recipients in Arkansas increased by 17.2 percent, welfare costs went up 60.1 percent, and total state spending went up 92.6 percent.

During the Clinton period, state taxes increased every year, sometimes by more than 10 percent a year. Other state taxes, such as sales-tax rates and gasoline-tax rates, also went up. The sales tax went up 50 percent from 1983 to 1992, from 3 percent to 4.5 percent, and the gasoline tax went up nearly 100 percent, from 9.5¢ a gallon to 18.5¢. So, although per capita income inched up toward the end of Clinton's era, his relentless increases in taxes added about $590 million per year (on the average) in tax burden to the Arkansas economy.

A couple of other variables showed promise near the end of that period, and others even improved, such as the unemployment rate, which fell from 7.6 percent in 1980 to 7.0 percent in 1992. But in context, the situation reminds me of the time my daughter was awarded the title "Most Improved Swimmer at Camp" back when she was in her early elementary school years. She swam no better than the rest of the kids, but she won the award because at the beginning of the summer, she'd declined even to get into the pool.

## A Plan to Create Eight Million New Jobs?

*"I've offered a comprehensive plan, a real plan to rebuild America, create eight million new jobs."*

> The New York Times
> quoting Bill Clinton
> September 17, 1992

Can any government plan really create jobs? And given this presumption, can it accurately predict how many jobs it will create? When politicians use big numbers to describe events in the past, we should be wary. But when they use big numbers to describe events in the future, we may as well switch to another channel.

Yes, it's certainly possible for the government to create eight million new jobs out of nowhere—say, by expanding the federal bureaucracy by eight million new positions. There's certainly a glorious tradition behind government job creation, dating back to the pharaohs' brave, pro-economic-growth pyramid-building schemes in ancient Egypt. And for a more modern example, we surely saw little or no unemployment in the Soviet Union. (No economic plan is perfect.)

Clearly, jobs for the sake of jobs is never a true measure of a healthy economy, and creating eight million jobs by federal fiat would not be without side effects. Again, let's look at the numbers on both sides of the ledger. On the one side, we now have eight million more jobs. But jobs imply salaries—and that's on the other side of the ledger. Where will the money come from?

The government has only three ways of making money: taxing, borrowing it, and literally printing it. But taxing takes the money out of the hands of citizens who would otherwise have used it to buy goods and services for themselves and their families. In addition, their spending supports jobs in the private sector. If the citizens no longer have that amount of money to spend, the relevant number of jobs will disappear. (Even simply *saving* the money makes capital available for business borrowing, growth, and subsequent employment.) It's impossible to quantify this effect precisely, but it's clear that the taxes to pay for those eight million new positions would cause a diminution of employment elsewhere in the economy, even though we can't always predict where.

The same is true of paying the salaries of the newly employed by borrowing, which limits the capital available for private investments, new businesses, home building, and all the other productive uses to which citizens put the money they borrow when the federal government doesn't crowd them out of the market for credit while in the pursuit of big numbers of new jobs for the sake of big numbers of new jobs.

But the third option is the one with the most insidious effects. By inflating the amount of currency in circulation (other things being equal), the value of each unit diminishes. This is the root cause of inflation, which works as a hidden tax, sapping the value of all Americans' earnings and well-being, in different, often unpredictable but still inexorable ways. Some people look at increasing the amount of money in circulation as a solution to all sorts of problems, but it always (and pervasively) interferes with the primary purpose of money, which is to serve as a stable unit of value. While its effect on jobs is uncertain, unchecked inflation can lead to systematic economic disaster—such as occurred in Germany in the 1920s—that far outweighs the ill effects of job losses alone.

Of course, Clinton didn't mean that his program would create jobs by directly employing eight million people for government projects. His plans for spending included $200 billion in government funds on "public investment," including high-tech infrastructure like high-speed trains, an "information superhighway," and improving education and job training. But regardless, all this government money—which is supposed to create new jobs indirectly, rather than through direct federal hiring—is *still* coming from either taxing, borrowing, or inflation. And tax money spent to create jobs building new high-speed railroads is still money taken from American citizens who might prefer to buy houses, automobiles, and computers for themselves, support our existing "low-speed" railroads, or, indeed, purchase any of a myriad of other goods and services that support another American's job.

Another major problem is that government usually directs its spending not on the basis of satisfying consumer desires and needs with the greatest efficiency but rather on satisfying the constituents in as many congressional districts as possible; in satisfying the needs of the bureaucracies supervising the spending; and in satisfying the political needs of the most powerful branches of the government. It's amazing how much a high-speed train can cost when all those different demands must be satisfied, rather than merely satisfying the desires and needs of consumers for transportation—desires and needs that are already very well met by automobiles, buses, planes, and trains that travel only twenty-five miles per hour less than the planned high-speed rail.

## Government Mandates Don't Cost Money?

*"'Clinton's economic program is going to cost us 2 million jobs.'... There are two fallacies in [this] reasoning. First [Senator Phil] Gramm [R-Texas] and other Republicans base these job-loss calculations on a false analogy: They treat a government mandate, such as requiring maternity leave or health insurance for employees, exactly like a tax.... From there calculating the alleged job loss is a mechanical exercise.... When the government tells businesses to do something, it is not taking anything from businesses and labor together ... it is terribly misleading to treat them as taxes."*

Business Week
article by Alan Blinder
September 21, 1992

Attempting to quantify the effects on one economic variable by measuring a change in another is never a mechanical exercise that can be trusted to give accurate result, so it is apparent that in one sense Blinder was correct in what he said in his article, "Clintonomics: Figure the Merits Along with the Math." There was no reason to take seriously the figure of "two million jobs" that Clinton's opponents were speculating would be eliminated.

But the principle behind the speculation is sound. Lost jobs can reasonably be associated with an economic program that involves employer mandates, and Blinder was looking at only one side of the ledger, even though he maintained otherwise. Blinder said that it was wrong to treat government mandates as taxes because with mandates, what is taken from one side—in this case, employers—is being given to another—in this case, employees. He saw this as no net loss. But this is also true of any *tax*, a comparison that Blinder explicitly denied. When government takes money from American taxpayers, it doesn't just burn it up. No, government will spend the money it receives, so someone, somewhere—whether a federal bureaucrat or a subsidized businessman or a welfare recipient—is always gaining in a tax transaction, in the same manner as someone is always gaining in a government-imposed mandate.

In the case of a tax, money is taken from one person and given to another, and in the case of a demand made upon an employer, the act of hiring and keeping an employee is made more expensive by the amount that the employee receives in value, such as maternity leave (that must be covered by expensive temporary help or by coworkers working longer hours or by forgoing some amount of production) or health insurance (that the company must purchase). While it is possible for those already employed to benefit from a mandate, this is not necessarily the case. Whether or not employees receive a benefit that they value, the consequence still remains that the increased employer expense only serves to hold salaries down and make hiring new workers more expensive.

Regardless, it is one of the basic tenets of economics that when something costs more (other things being equal), less of it will be demanded. But will this amount to two million jobs? Probably not.


### George Bush Was a Hypocrite?

*"If the Congress had appropriated every last dollar George Bush recommended, the deficit would be bigger today than it is. They talk about cutting spending. . . ."*
The New York Times
quoting Bill Clinton
September 21, 1992


Let's have a numerical look at our so-far-untarnished (at least, in this book) forty-first President of the United States, George Bush. It's no wonder that Clinton scoffed at his rival. In every area of federal government spending—except defense, where Bush reduced his requests for appropriations from 1989 to 1992—this President appeared never to have met a spending bill he didn't like. And he wasn't forced into it by a steamroller over-spending Democratic Congress, either. Bush not only never vetoed a budget because of spending increases, he actually asked for them himself. In his 1992 to 1993 budgets, while he was preparing for and participating in a campaign against Clinton, whom Republicans were denouncing as just another "tax-and-spend" Democrat, Bush was requesting the following spending increases:

> 16.1 percent for Housing and Urban Development
> 15.6 percent for the State Department
> 14.7 percent for the Department of Education
> 11.2 percent for the Treasury Department
> 10.6 percent for the Justice Department
> 7.6 percent for Health and Human Services
> 5.1 percent for the Environmental Protection Agency*

By now, assuming the reader is becoming more and more sensitive to numbers and what they may and/or may not mean, he or she should be wondering how Bush's spending compares with other Presidents, both Democrats and Republicans. As we discuss that, we should keep in mind that percentages become more "important" as their base becomes larger. That is, with a budget of $20, a 100 percent increase isn't significant; but with a budget of $20 billion, a 100 percent increase is certainly significant. So when comparing Bush's percentages to past presidential percentages, we must then keep in mind that most of today's budgets are larger than yesterday's. (Of course, we must also keep in mind the effects of inflation. The list continues. Suffice it to say that apparently simple numbers are far more complex than we might once have thought.) Now let's compare Bush's record to that of his immediate predecessors, especially "tax-and-spend" Democrats like Jimmy Carter and Lyndon Johnson.

Bush's average annual real domestic-spending increases from 1989 to 1992 were 8.7 percent. Carter managed to hold his spending increases to a comparatively thrifty 3.5 percent. Even Lyndon Johnson, the father of the Great Society—that notorious linchpin of big-spending Democratic social programs—hit only 5.5 percent. And he was subsequently dwarfed by his successor Richard Nixon, who challenged Bush for the big-spender title by reaching an average increase of 8.5 percent per year in real domestic spending. Casting a numerical net into the past, we must go all the way back to John F. Kennedy to find a President who managed to top Bush's average spending increases, but only just barely—Kennedy's were 9.0 percent.

*All percentages are in inflation-adjusted dollars.

Thus, Clinton caught Bush in a blatant hypocrisy about cutting spending. (But for the most literal of readers—yes, Bush never did criticize those "spending" Democrats; rather, he excoriated those "*tax*-and-spend" Democrats!)

### Capping Is the Same as Slicing?

*"A senior White House official said . . . he [Bush] favors putting a cap on all mandatory spending programs except Social Security so they could rise no more than the rate of inflation and the rate of increase in the eligible population. Such a cap, he said, would save the government $294 billion over the next five years. . . . [But] he offers no clue how it would work, nothing that would give any voter any indication that a favored benefit might be sliced."*

The New York Times
September 24, 1992

It is an article of continuing faith among scholars of deficit-cutting that the only way to achieve real reduction is through some combination of two means, each of which is effectively (meaning, politically) impossible—cutting spending and/or raising taxes. (Why is this impossible? Because politicians assume that the majority of voters are mathematically illiterate.) But if we look at the numbers beyond this belief, does it stand to reason?

The above quote refers to an approach to deficit-cutting that Bush promoted during his 1988 campaign—the notion of the "flexible freeze." In short, Bush stated that he would enforce a ceiling of 3 percent annual growth in total federal expenditures. That was the "freeze" part. The "flexible" part allowed for raising spending in certain areas above this 3 percent, but only if an offsetting decrease was made in another area. (That is, another area must increase less than 3 percent.)

Bush abandoned this idea once he became President, but if he had implemented it, it would have resulted in a balanced budget by 1993—without any new taxes. No spending cuts or tax hikes would have been necessary to achieve total deficit elimination—all that would be required would be holding spending and taxing approximately steady. Instead, Bush went along with the Budget Enforcement Act of 1990 instead, which featured a planned $160 billion in new taxes over five years. We shouldn't fault the skepticism of *The New York Times* when it looks askance at political claims involving such big numbers, but maybe we should fault it for wondering how such a spending cap would work, that is, what "favored benefit might be sliced." With a cap, nothing is sliced at all; instead, growth rates are slowed.

Not that slowing growth rates necessarily ends the problem in time to prevent a catastrophe. While we may claim that we're not gaining weight as quickly as we used to, this doesn't mean that we're losing weight. (The days of our growing from ten pounds to twenty pounds—a 100 percent increase—are past, but if we grow from two hundred pounds to three hundred pounds—"only" a 50 percent increase—is that tolerable?)

Slowing growth rates is a first step, relatively painless, and if kept under steady pressure, would work. If Bush had stayed with the flexible-freeze option (or returned to it, as the anonymous official was hinting might be the case in a second Bush term), federal spending in 1992 would have been $1.25 trillion instead of the $1.46 trillion it actually was. The deficit in that same year would have been a mere $87 billion as compared to the actual $277 billion.

### Everyone Should Earn More than Average?
*"The much-touted job gains of the 1980s were, for the most part, low-wage positions earning $250 a week or less. More than 25 percent of the U.S. work force now toils in this class of jobs."*

*Time*
September 28, 1992

For the moment, suppose these numbers about job gains are accurate. Is the high numbers of low-wage jobs a problem? Unless all wages are equal, aren't some always going to be low by comparison, some middle, and some high? Why wouldn't 25 percent lower-income jobs, 50 percent middle-income jobs, and 25 percent upper-income jobs be perfectly normal? The human population routinely graphs that way in plenty of other areas.

In a free-labor market, people receive wages that roughly reflect their worth to their employers. If they are paid less than that, competing employers can easily bid them away and still make a profit. If paid more, their employers will lose money and be forced to take remedial action or their businesses will fail. This is why public policy that increases employee remuneration—whether through higher minimum wages or employer mandates (like medical insurance)—will ultimately increase unemployment.

That aside, what's inherently wrong with a great many jobs opening for our least-educated, least-skilled, often-otherwise-impoverished citizens? Try as they might, not everyone can muster the capabilities necessary for middle-level jobs, and even fewer of us can handle high-level jobs that require managerial ability, technical expertise, or special talent. Or if, because of youth or inexperience, a low-paying "first job" is all a person can obtain, why should this offend us?

Also, value systems differ. The thought of staying in school beyond what's required, assuming much more responsibility, or spending day after day doing somersaults in order to get along with competitive co-workers and cranky bosses, all in the pursuit of the almighty dollar, is just plain intolerable to some people. Intellectually undemanding jobs suit some peoples' personalities just fine, and who's to say they have some sort of problem because of that? There are plenty of people who would like to ditch their briefcases if only they knew how to get back off the fast track.

And what about all the people whose mates make an acceptable living, and who want only to supplement that income a little bit and maybe keep busy and make a few friends at a fairly minimal job that will help them feel (and be) productive and keep them in touch with what's going on in the world? Moreover, what many people seek in a job is a significant degree of flexibility and personal freedom, often to take care of small children. Part-time jobs are perfect for them, but they're usually "low-wage," if for no other reason than that the work hours are fewer. All other things being equal, part-time jobs always pull down the average, and the average they pull down is always at the lower-income levels. There are no part-time jobs for people running major corporations.

At any rate, 82 percent of the jobs created in the United States from January 1982 to December 1989, according to Bureau of Labor Statistics, were in areas classified as "technical," "production," and "managerial/professional," with the average median yearly earnings among those three categories for 1989 being, according to the Census Bureau figures, $26,536. This median figure is more than *twice* the yearly income of someone earning the $250 per week mentioned earlier. (See Part Two for more about the concept of "median," which is often the "average" of choice when considering income levels.) Then again, if none of these jobs had opened, then at least the average American salary would go up, wouldn't it?!

### A Tax Break for Millionaires?

*"George Bush signed the second-biggest tax increase in American history . . . now George Bush wants to give a $108,000 tax break to millionaires. $108,000! Guess who's going to pay?"*

The New York Times
quoting Bill Clinton
October 3, 1992

With the help of his own case of numerical naïveté, Clinton struck a mathematical blow at his opponent and actually wound up *understating* the full case he could have made—at least with regard to the first half of his statement. It wasn't merely the second-highest tax increase in American history that Bush signed with the Budget Enforcement Act of 1990. The agreement also completely derailed deficit reduction from the steady downward path it had taken since the enactment of the Gramm-Rudman-Hollings Deficit Reduction Act of 1986.

The deficit was $244.6 billion when Gramm-Rudman was put into effect. By 1989, it had shrunk to $152.5 billion. (Both figures are expressed in constant 1989 dollars.) As a percentage of GNP, an important consideration in terms of the burden the deficit could be considered to place on the overall economy, the deficit had gone from 5.3 percent to 3.0 percent during the same period. But despite this progress, Bush found it necessary to negotiate a compromise with a hostile Congress to supersede Gramm-Rudman in 1990. At the time, this new agreement promised to reduce total budget deficits by around $500 billion over the course of the years 1991 to 1995.

It will surprise no one that shortly after the agreement was signed, some of the numerical realities became apparent, and these estimates needed to be revised. (Projections of the results of tax raises, and estimates of deficits that are dependent on such revenue projections, are constantly changing in official government statistics. In fact, even figures from the recent past can change when new documents are issued. Reports of spending from 1992 can change from 1993 to 1994. This provides another clue to the soundness of government figures.)

By then, estimates of total dollars in that same period showed that federal budget deficits, far from being reduced by $500 billion, would instead *increase* by $708 billion. Adding these numbers floodlights an error that is truly awesome in its scope, even in modern times. *The estimates were off by $1.2 trillion.* (This may give us pause when remembering that the federal government handles much of our retirement and health-care plans.)

In the course of acquiescing to this settlement, Bush thus enacted a tax increase—including the income-tax hike for only upper-income Americans which Congress wanted so badly—that was the largest in history, at least in its initial effects and its intentions, not merely the second-largest, as Clinton later stated.

Before getting to the likely source of Clinton's understatement, let us note that while the overall tax increase was calculated to raise revenues $160 billion *above* what they otherwise might have been during the 1990 to 1995 period, revised figures—with the passage of some of those projections into cold reality—indicate that tax revenues collected by the federal government will instead be $529 billion *less*. A likely reason for this shortfall was an unanticipated recession that the tax increase may well have helped bring about. For every $1 that this increase intended to be collected by the federal government, a little more than $3 was lost. Such is the counter-intuitive relationship between raising taxes and raising revenues.

The numerical distinction between the 1990 budget compromise as the largest tax increase or merely the second-largest hinges on whether we consider the effects of a tax increase during its first year—when the country begins to shoulder its weight—or we compare estimates of how much the tax increase will cost the country over the long term. When doing that comparison, Bush's 1990 Budget Enforcement Act competes with Reagan's similarly unbeloved 1982 Tax Equity and Fiscal Responsibility Act (TEFRA) debacle.

TEFRA was a mighty tax act, indeed. The first year after it was enacted, the total tax burden of the country was increased by $17.3 billion. If we adjust those 1983 dollars for inflation, translating them into 1991 dollars, we find that the 1998 tax increase of $17.3 billion translates into a 1991 tax increase of $23.7 billion. But even this does not match the first year of the 1990 budget deal. In 1991, it increased the total tax burden of the country by another $25.5 billion.

So how did Clinton miss this chance to deck his opponent for signing off on the biggest tax-guzzler in the history of the nation? Probably by looking at fiscal analyses that projected the long-term effects of the 1990 Budget Act versus those of TEFRA. While this isn't illegitimate—although one must be aware that initial effects can be very different from long-term ones—any multiyear conclusion must depend on a great many economic projections, and everyone knows the value of those.

But what of the second part of Clinton's claim—that Bush was planning a "$108,000 tax break to millionaires"? In fact, he had just raised their taxes. Then, while on the campaign trail, Bush had talked about a 1 percent cut in overall income-tax rates, which would give all citizens some tax relief. The Clinton team then calculated the portion of this relief that would go to millionaires as being worth $8,000 apiece on the average. The rest would come from the 12.6 percentage point capital gains rate tax cut proposed by Bush.

Those figures come from an analysis of 1990 taxpayers, which shows $40.4 billion in capital gains accruing to 40,385 millionaires. But what about all the rest of us capital gains taxpayers? As explained earlier, there are far more of *us* than there are of those millionaires, but Clinton conveniently ignored this fact. Moreover, his figures also rested on the assumption that a capital gains tax cut actually would result in lower revenues, which contradicts all historical experience.

Moreover, would Clinton have suggested that lower-income and middle-income taxpayers be furious if Bush had proffered a 50 percent income tax cut to all Americans? According to Clinton's faulty logic, because upper-income Americans pay so great a percentage of their incomes in tax, a tax cut would benefit them more, and therefore, lower- and middle-income people should reject it. Which means that only if upper-income people pay a very *small* percentage of their incomes in tax, should lower- and middle-income people accept a cut.

Finally, there's the presumptive, "Guess who's going to pay?" No one needed to pay anything. (If we have any doubt about it, we need to go back and reread the pages on logical fallacies.) Is it presumed that everything we earn is to be the property of the government and that every dollar we are "allowed to keep" is somehow taken from the government and, by illogical extension, from everyone else?

### Job-Seekers Should Move to Arkansas?

*"For 12 years, [Clinton] battled the odds in one of America's poorest states—and made steady progress. Arkansas is now first in the nation in job growth. Even Bush's own Secretary of Labor called job growth in Arkansas enormous. He moved 17,000 people from welfare to work. And he's kept taxes low. Arkansas has the second-lowest tax burden in the country."*

> The New York Times
> quoting campaign ad
> for Bill Clinton

The first statement made above was quite right as far as what the official numerical record showed at the time. From July 1991 to July 1992, and during the tail end of a national recession yet, Arkansas was blessed with a 3 percent rise in nonfarm employment, the fastest in the nation. (As a commentary on "official records," however—not about Arkansas in the Clinton era—newer numbers from the U.S. Bureau of Labor Statistics, issued on March 1, 1994, show Arkansas as having a 3.7 percent rise in employment in the 1991 to 1992 period, even higher than before, but now ranking it in third place, behind Montana with 4.22 percent and Idaho with 4.21. So much for the accuracy of "official records.")

But the quote, "Even Bush's own Secretary of Labor called job growth in Arkansas enormous," was out of context. Then-Labor Secretary Lynn Martin's full comment was "Arkansas' job growth is enormous, that's true, *if you're working from a low base.*" That is, when starting from a low base, a high percentage rise in number of jobs created doesn't necessarily translate into a high total number of new jobs created, though the latter number usually is more significant to people, depending on the population. (A state that started with only one job and mushroomed that base into three jobs would have a growth rate of 200 percent.) In the case of Arkansas, 38,000 new jobs were created from 1991 to 1992, but fifth-ranked Kansas created 41,000 jobs, and first-ranked Montana created only 14,000.

The ad's claim about moving 17,000 people from welfare to work was true on one level; but on another level, it was an audacious example of pointing to only one side of the ledger. Yes, according to officials at Arkansas' Department of Human Services, 17,000 people were indeed removed from food stamps or Aid to Families with Dependent Children (a major welfare program) and into jobs through a state government program from July 1989 to June 1992. But this was a gross figure, not a net one. In other words, how many people moved *onto* the welfare rolls? The answer is: plenty. The Aid to Families with Dependent Children program in Arkansas took on 3,334 new members, and households receiving food stamps added some 22,109 people, bringing the grand total to 25,443. Subtracting 25,443 from 17,000 leaves us with the net figure of 8,443 *added* to the welfare rolls.

(Also, what was that complaint about "the much-touted job gains of the 1980s" being "for the most part, low-wage positions earning $250 a week or less"? What about the jobs in Arkansas? Did these former welfare recipients take jobs as executives?)

Now let's consider Arkansas' tax burden. Was it really the second lowest in the country in the way that most people would understand this statement? When trying to get a mathematical grip on the claim—which is supported by measuring the per capita (per person) combination of state and local taxes—it is essential to remember that Arkansas is the fourth-poorest state in the nation; thus, its per capita tax burden could be comparably low. In other words, when people don't make much money, they don't pay much tax. To compensate for this deceptive distortion, let's consider Arkansas' tax burden per $1,000 of income instead. That is, when people *do* make money, how much tax do they pay? Now things look a little less good. Using this measure instead, Arkansas had only the *eighth* lowest state and local tax burden.

But we're not finished. If we're supposed to use the figures in the ad to judge the effect of *Clinton* on Arkansas' tax burden, then the combination of state *and local* taxes is inappropriate. It is only the *state* burden that can be considered Clinton's responsibility. He was the governor of Arkansas, not a mayor or a city council head of any of the state's many localities. So when the relatively lower local taxes—which were not Clinton's achievement—are factored out, we find that at $1,146 per capita, Arkansas had only the nineteenth lowest state tax burden.

And we're *still* not finished. That was the *per capita* tax burden—a misleading measure for Arkansas. So to compensate for this distortion, let's again consider the tax burden per $1,000 of personal income, the most relevant figure for what this ad purported to show. Now we discover that at $79.21 per $1,000 in 1992, Arkansas had the twelfth *highest* state tax burden in the country.

### Bush Refused to Increase Taxes?

*"The main reason the deficit has risen so much is that although government spending increased during the Bush presidency by an average of 7.5 percent a year, after adjusting for inflation (as against 1 percent under Ronald Reagan), Mr. Bush refused to support offsetting tax increases."*
The New York Times
October 5, 1992

When combined with these numbers, the use of the word "refused" is an effective logical twist.

During Bush's Administration, domestic spending grew at an inflation-adjusted annual rate of 8.7 percent—faster than it grew with any of his five immediate predecessors. (The discrepancy between this number and *The New York Times*'s 7.5 percent comes from factoring out the declining military budget.) From 1989 to 1992, almost every imaginable area of federal domestic spending rose by well over 10 percent in real terms. In some areas, the growth was staggering, such as the Departments of Commerce and State considered together, where appropriations rose 48 percent, or the Departments of Labor and Health and Human Services considered together, where appropriations rose 63 percent.

Bush's 1990 Budget Enforcement Act was intended to raise taxes by $160 billion. For 1992, when *The New York Times* made its comment, this would have brought federal taxes up to 19.5 percent of GDP, and for the period from 1990 to 1993, it represented a taxation level of 19.4 percent of GDP. No U.S. President since World War II has managed to create a tax burden as large as 19.4 to 19.5 percent of GDP. (Carter came closest, with 19.2 percent.)

Returning to the use of the word "refused," *should* numbers like this ever be supported with taxes? That is, why not maintain that the reason the deficit has risen so much is that *spending* increased so much?

## Our Wages Have Dropped From First to 13th?

*"We have gone from first to 13th in the world in wages in the last 12 years, since Mr. Bush and Mr. Reagan have been in."*

> The New York Times
> quoting Bill Clinton
> October 12, 1992

International wage comparisons like this are meaningless, because they don't consider all the practical factors that impinge upon them. For example, they don't adjust for parity in purchasing power, which is an adjustment calculation of what those wages can actually buy, given the differing price levels in the various countries. Also, the cost of living in most of the industrialized European countries, the area to which Clinton is referring, is far higher than ours. Moreover, wage comparisons don't adjust for the continuous exchange-rate fluctuations.

But even if we were able somehow to account for all these factors, there's a far more important consideration. Clinton ignored a significant side

effect of higher wages, and this effect holds true for higher prices in gen
eral. (Wages are, in fact, another term for the price of labor.) In a free
market, including a free labor market, if the price goes up, the demand
goes down.

This nearly universal law can be seen at work all over Europe, where
unemployment rates are much higher than in the United States, effectively
matching their so-called "higher wages." As of early 1994, average unemployment among European Community countries was 11 percent and
projected to be rising; at that same time, unemployment in the U.S. was
down to 6.5 percent. Since 1970, these European countries have only
managed to generate between three million and four million new private-
sector jobs collectively; during that same period, the U.S. generated forty-
one million. High wages also result in an inelastic market for labor—
during 1991, 46 percent of unemployed Europeans had been out of work
for more than a year; only 6 percent of the U.S. unemployed were in that
situation. Such are the wages of "high wages."

### Forty Million Americans Don't Have Health Insurance?

*"To define a health-care crisis, Mr. Gore claimed that '40 million' Americans 'who work full
time' do not have health insurance."*

> The Wall Street Journal
> op/ed by Alan Reynolds
> October 21, 1992

Gore can be forgiven for rounding up the estimates of uninsured Americans, which range from thirty-three million to thirty-seven million, representing around 15 percent of the population. However, that is the *total*
number of people without health insurance—not those without health
insurance who work full-time. According to figures from the U.S. Congress's Joint Economic Committee, half of the uninsured have full-time
jobs, lowering Gore's figure to 16.5 to 18.5 million instead of forty million.

But being uninsured is not an involuntary condition limited to those
who are poor. Only about 29 percent of the uninsured actually can't afford
medical coverage. In fact, 25 percent of the uninsured are young (from
ages eighteen to twenty-five), generally healthy men and women who simply choose to take the risk of paying their very infrequent health-care

expenses out-of-pocket rather than paying a monthly premium for insurance they feel they don't need. This is the carefree way of the young—they think nothing can ever happen to them. Just as their parents have trouble convincing them to keep good company and call every Sunday, so do their parents have trouble convincing them to spend their money on health insurance rather than on new clothes and entertainment.

So Gore ignores the numerous reasons—many of them perfectly logical, depending on one's value system—for people to decide not to buy health insurance. Some people may prefer a bigger apartment or a newer car or a more exotic vacation than they would be able to afford if they purchased health insurance first. They assume that health care will be available to them at their local hospital's emergency room if they ever really need it, which is generally true.

This is a matter of personal priorities. When I was a child, medical insurance ranked first among the families on our block. I don't know where it ranks on that same block now. Maybe child care for the working mothers has displaced it. Maybe it even ranks last, which wouldn't be too surprising; a great deal of free care is indeed available, including immunizations for children. But times do change, and priorities change with them.

Another implication hidden behind these numbers is that the uninsured are a hard-core group of people who are chronically in need throughout their lives. The reality is that their names change from month to month. Many are simply between jobs. In fact, 50 percent of all the uninsured people in this country remain that way for only four months, and only 15 percent of the uninsured remain that way for longer than two years.

## The First Decline in Industrial Production Ever?

*"In the first debate, Mr. Clinton said we are suffering 'the first decline in industrial production ever.' "*

> *The Wall Street Journal*
> op/ed by Alan Reynolds
> October 21, 1992

People sometimes overstate the facts in the heat of a debate, so Clinton can be forgiven this "first decline ever" exaggeration. Moreover, industrial production is often in decline during a recession, and even though the recession was actually over by the time this debate was held, the relevant economic statistics had not yet been released, so he had no way of knowing it. In fact, industrial manufacturing in the United States had grown by 2.8 percent between May of 1991 and the first debate in October of 1992.

However, because Clinton usually conflated the Reagan and Bush eras in his campaign statements, maybe that was the time period to which he was referring instead. But from 1980 to 1990 (before the Bush recession began), the GNP went up in inflation-adjusted terms both in total and per capita (that is, divided by the number of Americans to give a per person "share" of the national production). Although the GNP is a flawed measure, to be sure, it's the only means we have of describing such a huge and disparate production scattered among hundreds of thousands of businesses and millions of individuals. During this period, total GNP increased by 30 percent, and per capita GNP increased by 18 percent.

This was slightly lower than it had been during the three previous decades, but not indicative of an overall downturn in industrial production. A decline in the rate of growth is very different from a decline, a concept that politicians would prefer that voters not realize. (With the American public's current state of numerical illiteracy, politicians can have it both ways, tailoring the figures to each individual argument.) The Federal Reserve keeps track of total industrial production and of manufacturing industrial production, and both of those rose as well from the beginning of 1980 through September of 1990, shortly after the Bush recession began. During that period, total industrial production increased by 20 percent, and manufacturing industrial production increased by 36 percent.

Critics disdained the job growth of the 1980s as being merely in the service economy, which they feel comprises jobs that are inferior in social class compared to the industrial economy. This was the origin of the complaint that America was becoming a nation of "burger-flippers," the stereotypical service job. While this attitude may be true to those who are class-conscious, many service jobs are superior to industrial ones in ways that are significant to the people employed in them. Consider, for example, the benefits of working in the environment of an insurance office or a kitchen versus that of a vehicle assembly line. (It could be noted that service jobs also include bankers, stockbrokers, and other so-called "respectable" positions.) Regardless, during the period from 1980 to 1991, average hourly earnings in the service economy rose by 6.8 percent, compared to a lower 4.8 percent rise in the wages of manufacturing jobs.

But industrial production didn't fall compared to the rest of the economy. In 1980, manufacturing output accounted for a 21 percent share of the GNP; in 1989, that share had gone up to 23 percent. In fact, according to an index kept by the Census Bureau, industrial production went up in every division except primary metals, tobacco, leather, and mining, and in nearly every case, it had gone up by more than 10 percent. Small losses in total manufacturing employment can be seen, but when this occurs in tandem with increased production, the result is economically positive—more production with less labor. This is the entire purpose of an industrial economy in the first place.

### We're Working Harder For Less Money?

*"Most people are working harder for less money than they were earning 10 years ago. . . ."*
The Wall Street Journal
quoting Bill Clinton
op/ed by Alan Reynolds
October 21, 1992

As this was another comment made during a presidential debate, Clinton can be forgiven a certain numerical imprecision. But even so, we know what he was implying. The word "most" in general usage means at least "more than half," probably even "a lot more than half." Let's rest in the neighborhood of 75 percent.

So, were most people working harder? If Clinton meant "working more hours," he was simply wrong. The official numerical evidence indicated just the opposite. The Bureau of Labor Statistics plainly showed that the average weekly hours of nonagricultural workers was down to 34.4 in 1992, a modest decrease from 34.8 in 1982. If he meant that many people working in 1992 weren't working at all in 1982, that's certainly true. (Many of these people were women entering the work force for the first time.) But it's certainly not true that they were then earning less than they were before they started working at all.

And what does "less money" mean? The Census Bureau, albeit with its confounding habit of dividing the American population into quintiles (which can lead to the many numerically conceptual difficulties discussed earlier), states that from 1982 to 1989, after adjusting for inflation, the median income of families before taxes went up 12.5 percent, and the after-tax income per person rose 15.5 percent. In 1990, the average household income was up from 1980 in every quintile. Any way you look at it, it appears that income was gained during the 1980s.

As an aside, there are many complications that arise when measuring figures such as this, such as problems with the consumer price index used to adjust inflation and problems with the accuracy of quintile analysis to begin with, as discussed earlier. The spending data—as opposed to income data—even shows people in the lower quintiles spending far more than their reported income. For example, the 1989 figures from the Bureau of Labor Statistics show that for the lowest quintile, real per capita spending equaled 212 percent of real per capita income.

This can happen for a combination of reasons. Income figures ignore federal government payments such as food stamps and Medicaid, although both are clearly worth money. Also, the surveys don't ask questions about savings, rent payments from boarders, loans, or gifts. For example, some people in the lowest quintile are students supported by unreported gifts from their parents. (Although it may be inappropriate to call young, low-income students "poor," they do constitute a steady portion—although the names change from year to year—of the lowest quintile.) And of course, income from any underground economy goes unreported to the government.

An additional problem is that the federal economic surveys are probably less accurate overall than they used to be. Back in 1948, 95 percent of the population responded to government requests for information. By 1985, only 72 percent did.

# Seven

▼

## From 1992 to 1996 and Beyond

We have just called into serious question many of Clinton's economic statements, but it was he who used and abused and confused the voters with the most numerical nonsense—especially when employing numbers to lend a superficial appearance of objectivity to an underlying appeal for votes. Both Perot and Bush used numbers far more responsibly. However, that doesn't mean they deserved to be elected. Other factors are paramount, and only the American people can decide what those factors are.

In future elections, we will hear more of the same until we demand more of our politicians. Then maybe we won't need to compensate for our "rational ignorance"—a term used to describe the perfectly reasonable decision made by any one citizen to accept political rhetoric at face value rather than going through the laborious research required to verify it—mainly because we simply don't have that kind of time. And maybe we won't need to compensate for our chronic case of numerical naïveté.

Among the most important responsibilities of living in a democracy such as ours is to vote thoughtfully, and the power of logical thinking can help us achieve that very worthy goal. One of the biggest weaknesses of majority rule is that the majority may be wrong. Slavery in eighteenth- and nineteenth-century America taught us that. Imagine the chaos of a classroom that allowed students to *vote* on what would be considered the correct answer. Logical and concepts are critical both to success and to the price to be paid for that success. That price is vital to understand. Just as we must never forget the Holocaust, it is imperative that we never forget the experience of the Soviet Union. It proved that achievement of a goal—

economic and social equality—bears little relationship to satisfaction with that goal, once achieved. That is, we must be very careful what we wish for—because we may get it. And it is the power of logical thinking that enables good hearts to make good wishes.

# Sources for Statistics in Part Two

1,2    James M. Dabbs, Jr., Georgia State University (Atlanta, Georgia).
3    Food Marketing Institute (Washington, D.C.).
4    Harold Takooshian, Fordham University (New York, New York).
5    Krups North America (Closter, New Jersey).
6,7    National Pork Producers Council (Des Moines, Iowa).
8    *Guinness Book of Records*, 1992, Facts on File (New York, New York).
9    Gallup Organization (Toronto, Ontario, Canada).
10    Research USA (Chicago, Illinois).
11    *Esquire's* College Survey (New York, New York).
12    New York Mets/*Harper's* Research (New York, New York).
13,14    Allensbach Institute (Allensbach, Germany).
15    Peter D. Hart Research Associates (Washington, D.C.).
16    A.C. Nielsen Media Research (New York, New York)/National Football League (New York, New York).
17    Randy Larser, University of Michigan (Ann Arbor, Michigan).
18    Whiteshell Laboratories (Pinawa, Manitoba, Canada).
19    Heinz USA (Pittsburgh, Pennsylvania).
20    Associated Press (Taipei, Taiwan).
21,22    *Time*-CNN Poll, by Yankelovich Partners (Westport, Connecticut).
23    National Wild Turkey Federation (Edgefield, South Carolina).

# A p p e n d i x : The Monty Hall Dilemma

## To Switch or Not to Switch
by Donald Granberg
University of Missouri

*A very bright gal named Savant*
*Made a game show mistake quite flagrant,*
*Ph.D.s by the throng*
*Wrote to tell her she's wrong,*
*But she wouldn't 'fess up or recant*

William J. Cohagan
Sun City, California

*The Monty Hall Dilemma Stirs Up Controversy**

In the September 9, 1990, issue of *Parade*, Craig F. Whitaker of Columbia, Maryland, posed this query to Marilyn vos Savant:

> Suppose you're on a game show and you're given the choice of three doors. Behind one door is a car; behind the others, goats. You pick a door, say number 1, and the host, who knows what's behind the doors, opens another door, say number 3, which has a goat. He then says to you, "Do you want to switch to door number 2?" Is it to your advantage to switch your choice?

Vos Savant's answer was direct and unambiguous,

> Yes; you should switch. The first door has a one-third chance of winning, but the second door has a two-thirds chance. Here's a good way to visualize what happened. Sup-

---

*A more complete description of the studies we have done, as well as complete list of citations, are available in D. Granberg and T. Brown, "The Monty Hall Dilemma," *Personality and Social Psychology Bulletin*, 1995, *vol. 21*, pp. 711–723.

pose there are a *million* doors, and you pick door number 1. Then the host, who knows what's behind the doors and will always avoid the one with the prize, opens them all except door number 777,777. You'd switch to that door pretty fast, wouldn't you?

Despite her explanation, she received a large volume of mail, much of which was from irate and incredulous readers who took issue with her answer. Vos Savant held her ground and cogently defended her answer in three subsequent columns devoted to this topic.

This problem has been dubbed the "Monty Hall Dilemma," after the host of the game show, *Let's Make a Deal,* which featured two-stage decisions involving options to stick with one's initial guess or switch to another alternative. By now, it has been discussed in considerable detail in the *American Statistician, American Mathematical Monthly, Mathematical Scientist, Mathematics Teacher, Skeptical Inquirer, The New York Times,* and elsewhere. The consensus is that Marilyn's answer was *essentially correct,* provided one makes some highly plausible assumptions:

A. The car is placed randomly with each door having an equal chance of containing the car.
B. The host knows where the car is and also knows what door the contestant selected initially.
C. The host uses that knowledge to select an unchosen door with a goat to show to the contestant after the initial guess.
D. The host is committed to showing an incorrect, unchosen alternative deliberately after the initial guess.
E. If the host has a choice between which of two incorrect doors to show, i.e., when the contestant's initial guess is correct, the choice is made randomly with equal probability.
F. The host is committed to giving the contestant the choice of whether to stick or switch.
G. The host is truthful.

With two losing doors and one winning door, a *knowledgeable host* can always show one losing door, regardless of which door is initially chosen. Thus, when the host *knowingly* opens an incorrect, unchosen door, this gives no additional information about the likelihood of the initially chosen door being correct. So under these assumptions, the probability of the initially chosen door being correct remains unchanged at .33. The probability that one of the unchosen doors contained the prize was .67 and this also remains unchanged when the host knowingly opens one of the unchosen doors to show that it is a loser. In effect, the host is offering a choice between the initially chosen door and the other two.

If one starts tampering with these assumptions, the situation can change rapidly. The wording of the question, as posed to vos Savant, was entirely compatible with these assumptions with at most some minor ambiguity on a few points. Most people who wrote letters disagreeing with Marilyn's solution objected for other invalid reasons and were therefore simply wrong. Letters came from a great variety of people and places, from people with lofty titles and affiliations and from others of more humble circumstances. For example, Andrew Bremner of the Department of Pure Mathematics at Cambridge University in England, wrote with a touch of *noblesse oblige*.

▼▼

Dear Marilyn,
    . . . your answer that you should switch to door number 2 . . . is incorrect. Each of doors number 1 and number 2 has 1/2 chance of winning. . . . Your correspondents seem rather rude; I wager your womanhood is a factor!
                                        Yours sincerely,
                                        Andrew Bremner

A somewhat more strident tone is evident in this letter from a professor of mathematics at Florida State University:

▼▼

Dear Ms. vos Savant:
    It is apparent from your "Ask Marilyn" column, dealing with probabilities, . . . that being smart is no guarantee of being correct. Your analysis of the game-show probabilities, and the analogy involving the pea under a shell, reveals a misunderstanding of the rudiments of probability theory, and an appalling lack of logic.
    . . . The probabilities that we assign to a set of future events are a measure of our ignorance. As we gain information about one event of the set, the probabilities that we assign to *each* of the events will change. . . .

As the contestant's knowledge of specific events increases, the entire set of probabilities changes.

Consider first the shell game, in which a single pea is placed under one of three shells. Let's label the shells A, B, and C and suppose that one player, named Abe, puts his finger on shell A and a second player, named Ben, puts his finger on shell B. In the absence of any information of the location of the pea (other than the facts that there is only one pea and it is under one of the three shells), we assign a value of 1/3 to the probability of the pea being under any one shell: $pA=pB=pC=1/3$. (Note that the probabilities must sum to 1.) Now suppose that shell C is lifted and the pea is not seen. We now know that $pC=0$. According to your argument, now $pA=1/3$ and $pB=2/3$ because "nothing has been learned to allow us to revise the odds" on the shell under Abe's finger. But by the same reasoning, $pB$ should remain 1/3! Now either $1/3 + 1/3 = 1$ or your argument is wrong. I leave it to you to figure out which.

. . . I urge you to lower your mantle of omniscience and (following the lead of Ann Landers) seek the advice of experts when the subject matter is outside your area of expertise. Your ignorant responses are hurting the fight against mathematical illiteracy.

Sincerely,
David Loper
Professor of Mathematics

Loper's extension of the situation to two players fundamentally alters the problem in that the host no longer has a degree of freedom over which door to open. In the original problem, no matter which door the contestant chose, the host could *knowingly* open some other door to reveal a goat. That is no longer the case with Loper's two player problem, and therefore it is a different problem. While Professor Loper suggested that vos Savant should "seek the advice of experts," other writers went even farther, suggesting that her column be discontinued. From State College, Pennsylvania, came this rather hostile letter:

▼▼

Dear Marilyn,

. . . As an educator, I find your 2 December column reprehensible, and urge you to reconsider your future endeavors. In my opinion, you no longer provide a useful service to your readers . . . I believe your 2 December article has discredited your reputation and, accordingly, I am encouraging *Parade Magazine* to drop your column. I would also encourage your publisher to print a strong apology to the three scholars whose wisdom was impugned by your illogical babble. I hope they sue you!

Most sincerely,

Gregory S. Forbes

Or consider this letter from a mathematician who listed his credentials as including a B.S. and Ph.D in Math at M.I.T. and Brown, respectively, more than thirty published articles plus a graduate text, faculty positions at Brown, University of Arizona, University of Southern California etc., and consultant at Los Alamos, the Jet Propulsion Laboratory and elsewhere:

▼▼

Editor, *Parade*:

My mathematics credentials are listed at the end of this letter. In September, Marilyn incorrectly answered a question referring to the probability of a prize behind one of three closed doors. Several professors of mathematics, including me, wrote to her, noting the error. I have on my shelf more than 50 probability texts, each of which solves the problem correctly, and warns against Marilyn's error, as a result frequently obtained by unskilled mathematical amateurs.

Today, Marilyn devoted an entire column to the problem, and included some of the critical letters. However, instead of graciously acknowledging her mistake, she ridiculed the the comments of the professors, and incredibly persisted with the myopic, convoluted logic which led to her original untenable result. Her defense of a known error today damaged *Parade*'s reputation for the truth. Her column should be discontinued.

Sincerely,

Paul Slepian

Nor were strong feelings limited to those who disagreed with Marilyn's two-thirds solution. Consider this letter from a professor of applied mathematics at Babson College:

▼▼

Dear Ms. Savant,

. . . Your solution is the correct one and any REAL mathematician can produce a proof of its correctness. REAL mathematicians consider this a trivial problem. . . . WHAT discipline do these respondents have their Ph.D.s in? Is it adolescent behavior? If it is in mathematics, my second question is what institution granted it? Really, their behavior is so disgusting I have lost sleep over it. . . .

Good luck and good wishes from Massachusetts.

Sincerely,

Stephen J. Turner

Several of the letter writers were so sure that Marilyn's analysis of the game show problem was wrong that they offered to wager large sums of money, e.g. $20,000, in a situation that would test whether one could truly gain an edge by switching in the Monty Hall Dilemma. Others wrote in a lighter vein. For instance, consider this brief letter:

▼▼

February 19, 1991

Dear Marilyn,

I have been following the debate about the game show problem with great joy. It is heartening to know that the professors and citizens of your great country can't see the error of their ways. If this is an example of American genius, then I will be victorious. HAHAHAHAHAHA!!!!!

Sincerely,

Saddam Hussein

Random Bunkers

Somewhere in Iraq

It is, of course, highly improbable that this letter was written by Saddam, especially in light of the fact that it was postmarked from Boston, and the return address on the envelope indicated it was sent by Robert Scheinerman, Suzanne Reyes, and Susanto Basu of the Department of Economics at Harvard University.

*Monty Hall Dilemma as a Research Problem*

It occurred to my colleague, Thad Brown, and me, that the Monty Hall Dilemma might have some potential as an intriguing research problem for cognitively oriented social psychologists. Our interest was not in the mathematical solution to the problem *per se*. Rather, for us the crucial questions concern what people experience, how they behave, and the process by which people reach a decision when faced with a two-stage dilemma like this. That is, is the solution to this problem really counter-intuitive and if so, why? Are there psychological blinders at work that prevent people from getting it correct? Are there relevant cues that are being ignored? Can people learn the solution inductively by playing the game repeatedly? In making two-stage decisions, do people become more committed to a tentative decision that they ought? These are among the questions we have begun to explore in the context of studying how people make decisions in two stages.

Of course, it is well established that humans are not objective information processors, and they are sometimes wrong in their intuitions about statistics and probability. A good example is the birthday problem, estimating the minimum number of people who must be in a room for the probability to reach .50 that two or more of them will have the same birthday. The correct answer of 23 is by no means obvious. Intuition tells us it must be more than 23. In one college class, 88 percent (50/57) made guesses larger than 23. The guesses averaged 184.

People use a variety of cognitive heuristics when acting as intuitive statisticians, and these heuristics can produce erroneous judgments. People are especially prone to error when faced with problems involving conditional probability and that is what is being considered in the Monty Hall Dilemma. Anecdotal evidence suggests that people will tend to *stick* with their initial hunch in the Monty Hall Dilemma when they ought to *switch*. Exactly why they would do this is not self-evident, but it may comprise a basic psychological tendency with considerable generality.

Consider the intuitive solution to the Monty Hall Dilemma. It may be thought that the probability of .33, located with the alternative shown to be incorrect after the initial decision, transfers in equal parts to the two remaining alternatives which would then each have a probability of .50. Inertia, commitment, or the desire to avoid being incorrect after switching may prevail, leads to the inclination to stay with the initial choice. If people didn't misapprehend the probabilities in this way, a different course of action might ensue.

In her four columns on this topic, vos Savant made one *minor* error; in her column of February 17, 1991, she summarized the responses she had been receiving, "Overall, nine out of ten *readers* completely disagree with my reply (*emphasis* added)." Mark Glickman of the Department of Statistics of Harvard University wrote to point out:

> If a reader agrees with your solution, the chance the reader will respond is small since there is no apparent controversy. On the other hand, if a reader does not agree with your solution, there will be a greater inclination to write in and express this sentiment.

The point that the *letter writers* could not be assumed to be representative of *readers* in the sense of comprising a random or representative sample is technically correct. However, it is an open question as to just how unrepresentative or atypical the responses of the letter writers are.

Ideally, one might like to include a Monty Hall type of problem in a Gallup Omnibus Poll so it could be answered by a representative sample of U.S. adults. Thus far, that has not been done. In laboratory experiments done at the University of Missouri, undergraduate students are shown three doors as in the Monty Hall Dilemma. After their initial choice and after being shown that one of the other doors is incorrect, they are given the choice of whether to stick or switch. In their initial encounter with the dilemma, 174 of 194 (90 percent) decided to stick even though their chances of winning would have increased to 2/3 by switching.

The Monty Hall Dilemma was also posed to 228 other undergraduates in a different format, as part of a survey completed in class. The wording, taken almost *verbatim* from John Tierney's article in *The New York Times,* leaves no doubt about the procedure to be followed by the host:

> Monty Hall, a thoroughly honest game-show host, has placed a new car behind one of three doors. There is a goat behind each of the other doors. "First, you point toward a door," he says. "Then I'll open one of the other doors to reveal a goat. After I've shown you the goat, you make your

final choice, and you win whatever is behind that door. You begin by pointing to door number 1. Monty then shows you that door 3 has a goat. What would your final choice be?

_____Stick with door 1_____Switch to door 2

In this study, only 13 percent checked "Switch to door 2." The gender difference was not significant; 15 percent of the men and 12 percent of the women said they would switch. Also, the switchers did not report higher grade point averages than those who said they would stick. While these 422 college students do not comprise a random sample of humanity or of U.S. adults, they were not self-selected in the manner of the letter writers; moreover they were naïve with regard to the Monty Hall Dilemma, and did not choose to participate because of their opinion about the dilemma. This evidence confirms the counter-intuitive feature of the Monty Hall Dilemma, and indicates that the reactions of the letter writers may have been, in this instance, rather similar to those of the readers.

In the laboratory experiments we also examined the trend across time by expanding it to a fifty trial situation. Thanks to the programming skills of Michael Hess, our experiment is administered and the results recorded by a computer. On each trial, the computer randomly assigns one door to be the winner. The subject makes an initial guess. Then the computer, acting in the role of the *knowledgeable* host, informs the subject that one of the other doors is incorrect; next the subject decides whether to stick with the initial guess or switch to the only remaining door. The computer then shows the subject which door is correct for that trial, and after being informed about winning or losing on that trial, the subject is ready to proceed with the next trial and so forth through fifty two-stage decisions. Subjects are told they will win $25 if their point total is the highest among the people in that experimental condition. We are finding that some inductive learning occurs across trials. Switching starts at a very low level, shows some increase, but then seems to reach a plateau of about 55 percent switching on the last ten trials.

We didn't know exactly what to expect when we began these experiments, but we did expect most people to stick initially; we were also not sure that switching would increase much across trials in our baseline condition. Therefore, we added two conditions where people gained double or quadruple points for winning by switching as compared to what they gained for winning by sticking. Under these conditions, switching began and finished at significantly higher levels. So people can be induced into a high level of switching if the incentive is strong enough.

▼

*Monty Hall and Russian Roulette Dilemmas*

The tendency of people to think intuitively they should stick with an initial decision when, in fact, they should switch, was dramatically demonstrated by the "Ask Marilyn" letter writers and by the subjects in our initial studies. We next wondered what people would do in a comparable situation in which it is rational to stick. Would people also stick when they should stick? Or, more perversely, would they switch when they should stick? To address this, we had to devise a situation in which it is rational to stick just as it is rational to switch in the Monty Hall Dilemma.

In the Monty Hall Dilemma, with one car and two goats, it is established that following an initial selection and the deliberate showing of an incorrect, unchosen alternative, one should abandon the initial selection and switch to the remaining alternative. However, what if we invert the problem so there are two cars and only one goat? In this version, the contestant would choose among three doors, two of which are winners and only one of which is a loser. We call this version the Russian Roulette Dilemma. (Inverted Monty Hall seems too cumbersome.) In Russian Roulette, there is usually just one bad chamber with a bullet. The rest of the chambers are all winners, i.e. empty. In words, the Roulette Dilemma can be described as follows:

▼

Suppose you are on a game show. The host, who is known to be completely honest, has placed a new car behind two of three doors. There is a goat behind the other door. "First, you point toward a door," the host says. "Then I will open one of the other doors to reveal a car and that door will no longer be available. After I have shown you that car, you make your final choice, and you win whatever is behind that door." You begin by pointing to door number 1. The host then shows you that door 3 has a car. What would your final choice be? Would you stick with door 1 or switch to door 2?

Under a set of assumptions which parallel those given previously for the Monty Hall Dilemma, in this situation it is rational to stick with one's initial decision. The probability of one's initial guess being a car is .67 and this is in no way altered by a host *knowingly* opening one of the other doors to show a car after you make your initial decision. Thus, in the three door Roulette Dilemma, the chance of winning by switching is only 1/3, while the chances of winning by sticking with one's initial guess are 2/3.

A close reading of the early literature on this topic made it easy to derive the Roulette Dilemma. The Monty Hall Dilemma is structurally equivalent, at least in terms of probabilities, to the Three Prisoner Problem presented by Martin Gardner in his "Mathematical Games" column in October 1959, *Scientific American*. In the three prisoner problem, one prisoner is going to be pardoned and two are going to be executed. The prisoners do not know their fates, but the warden does. Prisoner A asks the warden to tell A the name of one of the other men who will be executed. The warden complies and says that Prisoner B will be executed. A is glad, mistakenly thinking that his chances of being the one to be pardoned have just increased from 1/3 to 1/2. It is a small matter to invert the prospects so that two of the prisoners get pardoned and only one is executed. In fact, Gardner did just that when he discussed a situation in which the chances of avoiding death were 51/52. What is distinctive about the Monty Hall and Russian Roulette Dilemmas, in comparison to the Three Prisoner Problem, is the possibility of switching or sticking in a two-stage decision format.

Using both the Monty Hall and Roulette Dilemmas, we devised a two-stage decision-making situation in which it is rational to switch and a comparable situation in which it is rational to stick with one's initial choice. We have used the Roulette Dilemma with the Monty Hall Dilemma in two experiments. In a word problem study, we found that 15 percent said they would switch in the Monty Hall Dilemma, compared to 16 percent in the Roulette Dilemma saying they would switch. In that study, it appeared that people stick when they should stick and stick when they should switch. This implies a certain "stickiness" in two-stage decisions in which people become more committed to an initial decision than they ought to be.

We also did a laboratory experiment involving 50 trials of either the Roulette Dilemma or the Monty Hall Dilemma with the computer pro-

viding feedback after each trial and the prospect of a $25 prize. On trial
1 of that study, 8 percent switched in the Monty Hall Dilemma and 31
percent switched in the Roulette Dilemma. These percentages are significantly different from each other and they are also each significantly less
than 50 percent. Inductive learning did occur across trials, at least to the
extent of significantly differentiating these two conditions. In the final ten
trials, subjects in the Monty Hall Dilemma switched on 58 percent of the
trials, compared to only 26 percent switching in the Roulette Dilemma.

The implication is that the human brain is not wired to decipher readily
the rational solution in the Monty Hall and Roulette Dilemmas. On the
other hand, people can learn to differentiate them in a rational manner,
at least to a significant extent, if they interact in a given dilemma across
a number of trials. These two dilemmas, though perfectly symmetrical in
structure and logic, do not evoke responses that are symmetrical. In the
Monty Hall Dilemma where one should switch, people begin at a very
low level of switching but show some trend toward increased switching
across trials. In the Roulette Dilemma, where one should not switch,
switching begins at a relatively low level but there is no trend toward less
switching across trials.

Stated differently, the Roulette subjects were quite close to a correct
solution on the beginning trials but came no closer through experience.
The Monty Hall subjects were not close to a correct solution on the
opening trials but came somewhat closer through experience. If we invoke
a tough standard for complete mastery of the problem, namely switching
on all of the last ten trials in the Monty Hall Dilemma and sticking on
all of the last ten trials in the Roulette Dilemma, only 13 percent of the
Monty Hall subjects and 20 percent of the Roulette subjects met this strict
criterion. From this, it is safe to say that these dilemmas are not readily
learned inductively.

▼

### Knowledgeable and Agnostic Hosts

Joseph Moder of the University of Miami posed this delightful question:

▼

> In the Three Door Game Show, suppose the contestant
> picks a door as usual, but as the host reaches to open one
> of the other doors, he suddenly realizes that he has neglected
> to learn where the prize is. In the spirit of "the show must

go on," he unhesitatingly flings open one of the other two doors, and much to his relief, it does not expose the prize. No one, except the host himself, knew of the impetuous act he had just performed; the show went on *exactly* as it always had. Now . . . can the contestant improve his chances by switching doors?

---

Strange as it may seem, the answer is no. As indicated previously, the 2/3 solution to the Monty Hall Dilemma depends upon the host knowing where the prize is and also knowing the contestant's initial pick. Vos Savant's wording made the host's knowledge of the prize's location explicit and left us to presume the host would know the guest's initial selection and would use that knowledge in order to assure that a goat would be revealed behind one of the unchosen doors. An agnostic (without knowledge) host could do what appears to be exactly the same thing, e.g., show door number 3 with a goat and then offer the choice to stick or switch, but if the selection by the host was based on chance or luck rather than knowledge, the effect is different.

Consider these two scenes:

A. A deck of cards is placed face down in a random array on a table. You guess which one is the ace of spades by putting your finger on a card without turning it over. Taking this guess into account, the host, who knows where the ace of spades is, proceeds to turn over fifty cards, none of which is the ace of spades. The host then gives you a choice between sticking with your initial guess or switching to the only other card which is still face down.

B. A deck of cards is placed face down in a random array on a table. You guess which one is the ace of spades by putting your finger on a card without turning it over. The host then invites fifty other people, none of whom knows the location of the ace of spades, to each turn over one card. By chance, none of the cards turned over is the ace of spades. The host then gives you a choice between sticking with your initial guess or switching to the only other card which is still face down.

Under scene A, it is rational to switch, provided the other assumptions of the Monty Hall Dilemma prevail, since the chances of winning by switching are 51/52, compared to a chance of only 1/52 by sticking. On the other hand, under scene B, it does not matter what one decides, since the odds are even at 50:50 regardless of whether one decides to switch or stick.

Jeffrey Hoyt of the Department of Mechanical Engineering of Washington State University wrote to disagree with Marilyn's 2/3 solution to the Monty Hall Dilemma and offered this metaphor:

> Imagine three runners in the 100 meter dash. The runners are so evenly matched that the odds of any given runner winning are completely random. I am asked to guess the winner and select runner number 1. After making my choice, runner number 3 pulls a hamstring and cannot compete. Based on your answer to the game show problem, I would conclude that the unfortunate injury to runner number 3 has somehow inexplicably made runner number 2 the clear favorite over my original choice. By way of the above example, I hope it is now clear that your analysis of the game show problem is flawed.

A similar analogy was sent by Henry Seibel of Greer, South Carolina:

> ... picture a horse race with three horses, numbers 1, 2 and 3. Other things being equal, if horse number 3 drops dead 50 feet into the one mile race, the chances of each of the remaining horses winning the race are no longer 1 in 3 but are now 1 in 2. If this scenario is repeated enough times ..., then horse number 1 and horse number 2 will have a near equal number of victories.

These writers are correct in a limited sense, but the analogy to the Monty Hall Dilemma is fatally flawed by omission of a key feature of the Monty Hall Dilemma, namely, the knowledgeable host. An alternative wording might be:

> Suppose you have bet on horse number 1 in a 3 horse race in which there is no favorite. After you have placed your

bet, omniscient God who, of course, knows the horse destined to win and how you have bet tells you, "It's not going to be horse number 3."

Depending on God's other attributes, e.g., whether God seeks to guide people toward the right decisions, you should probably switch to horse number 2 if you have the opportunity.

The importance of the *knowledgeable* host is not apparent to some people who agree with vos Savant's 2/3 solution and give otherwise clever and insightful comments. For instance, James Simmons, a mathematician at the University of Virginia wrote:

I was amused by the elaborate attempts to explain the best strategy to use in the game-show contest . . . ("It wasn't until I started writing a computer program. . . .") If I am the contestant and my strategy is always to stick with my initial choice, then the probability that I *win* is 1/3. If my strategy is always to switch, then the probability that I *lose* is 1/3, since I lose only if the car is behind the door I first choose. As I either win or lose every time I play, the probability that I win using the switching strategy is 2/3.

Simmons's solution is correct, but only provided we assume the host to be knowledgeable and that the host uses that knowledge to expose a goat behind one of the unchosen doors after the initial selection.

Many people writing to vos Savant concerning the Monty Hall Dilemma reported having done a computer simulation and sent along a printout or a computer disk. An especially valuable one was sent by Joseph Heiser of the Mathematics Department of the Pemberton, New Jersey, High School. It included two simulations, building into them the assumption of the knowledgeable or the agnostic host, and demonstrated clearly the importance of this distinction.

If the question is why do most people initially get the Monty Hall Dilemma wrong, the first answer is that they misapprehend the probabilities in the situation. The reason why they do this is first and foremost, because they fail to take into account the importance of the knowledge of the host and how that knowledge is used. In one of our word problem experiments, the dilemma was posed as involving a knowledgeable host to eighty-eight people and to eighty-three people as involving an agnostic host. On rational grounds, the eighty-eight people should say switch, and the eighty-three people should say, in effect, flip a coin since it's a 50-50 proposition. However, in the knowledgeable host condition where it is rational to switch, only 15 percent said they would switch and in the agnostic host condition, only 10 percent said switch. These percentages are far from a rational standard and are not significantly different from each other. The indication is that people fail to take into account adequately the importance of the cue of the knowledgeable host. With or without a knowledgeable host, after estimating the odds at 50-50, it is an interesting psychological question as to why people strongly prefer to stick since they could do just as well by switching.

In that study, we also probed beyond the simple stick or switch decision and asked people to estimate the probabilities along the way and to explain their reasoning. In the Monty Hall condition with a knowledgeable host, by far the most common pattern was to estimate their probability of winning at .33 initially and .50 in the final stage. Many subjects indicated their impression that the probability was .50 on the final choice, and subscribed in one way or another to the idea that one's first thought is generally correct. In another part of the survey, people were asked to estimate the percentage of times when they changed their answers on multiple choice tests that the change was from a right to a wrong answer, from wrong to right, and from wrong to wrong. The average estimate of changing from right to wrong was significantly larger than the estimate of changing from wrong to right. In fact, research by educational psychologists indicates that changes from wrong to right on multiple choice tests outnumber changes from right to wrong by a ratio of more than 2:1.

It also deserves mention that the knowledgeable and agnostic hosts do not exhaust the possibilities. A few of the people writing to vos Savant claimed that the wording of the Monty Hall problem, as posed to her, was ambiguous since we are not told the motives of the host. There is some truth in that assertion. However, it does not follow that the laws of probability apply only when the host selects a door at random. The laws of probability also apply to the situation involving a knowledgeable host so long as the host is committed to using that knowledge to open an unchosen, incorrect door and to providing the option to switch or stick. Under these conditions, the probabilities suggested by vos Savant (.67 for winning by switching and .33 for winning by sticking) prevail. Tierney's rewording of the problem, given previously, fixes the host's strategy by stating in advance what the host will do and by assuring that the host is honest. His wording has the host saying, ". . . Then I'll open one of the other doors to reveal a goat." This means that the host *will* do it, not that he might or could. The key here is that the host's strategy and procedure are fixed in advance and therefore, not a matter of the whim, personality, or motivation of the host.

If the host has discretion to decide whether to open an unchosen incorrect door and whether to offer the choice of switching or sticking, then the situation becomes complicated. In that case, knowing the motivation of the host would be of value and ought to enter as one of many factors in deciding whether to stick or switch. At the extremes of motivation, consider the cases of the stingy, *devious* host and the *generous,* helpful host.

The devious host tries to minimize costs and the chances of the contestant winning, and consequently only offers the option of switching when the contestant's initial guess is correct. Of course, if this became known, the smart contestant would never switch, and her probability of winning when the option to switch is offered would be 1.0 by sticking and .00 by switching. The option to switch would become irrelevant, and the overall chance of winning would be 1/3.

The generous host is eager to help the contestant find the car and only provides the option to switch when the contestant's initial guess is incorrect. Similarly, if this became known, the smart contestant would always switch since the probability of winning when the option is offered would be 1.0 by switching and .00 by sticking. With a generous host whose procedure is known, the contestant wins every time.

Between these two extremes is a multitude of motivational orientations for the host. For instance, the *indifferent* host flips a coin to decide whether to proceed to the second stage in which an incorrect unchosen door is knowingly opened and the guest is given the choice of whether to stick or switch. When an indifferent host offers the choice to switch or stick, one wins with a probability of .67 by switching and .33 by sticking.

It is important to emphasize, however, that the host's motives and preferences are *irrelevant* when the host is following a fixed procedure and has no discretion on what to do. The assumptions stated at the outset stipulate that the host is not free to decide whether to show an incorrect alternative and whether to offer the chance to switch.

In this section, the focus has been on the characteristics (knowledge, motives) of the host. A similar analysis could be made of the contestant and the importance of her knowledge. In her third column on the Monty Hall problem, vos Savant pointed to a fascinating feature which can be called the *Martian Paradox*.

▼

> The original answer [switch and win with .67 probability] is still correct, and the key to it lies in the question: *Should you switch?* Suppose we pause at that point, and a UFO settles down onto the stage. A little green woman emerges, and the host asks her to point to one of the two unopened doors. The chances that *she'll* randomly choose the one with the prize are 1/2. But that's because she lacks the advantage the *original* contestant had—the help of the host.

What is so paradoxical about this is that the same arrangement of two doors has different probabilities for different people as a function of their different experience and knowledge. For the contestant who knows what door she picked originally, and what door was subsequently eliminated by the host, the probability of winning can be .67 if she switches. For the Martian who came on the scene when there were only two alternatives left and is not informed about what has happened, the two doors are equally likely at .50 to contain the car. R. H. Good, a physicist at California State University, Hayward, wrote an insightful letter about many facets of the Monty Hall Dilemma, and then concluded that it "seems almost as paradoxical as quantum mechanics."

▼

### Counter-factual Thinking

A counterfactual is an imagined alternative to an actual event, e.g., ". . . if only I had studied more in high school." In commenting on their reasoning in the study, several people showed signs of counter-factual thinking. For instance, one person wrote, "I wouldn't want to pick the other door because if I was wrong I would be more pissed off than if I stayed with the second door and lost." Another person stated, "Never change an answer because if you do and get it wrong it is a much worse feeling." Comments by two others were similar: "It was my first instinctive choice and if I was wrong, oh well. But if I switched and was wrong it would be that much worse." "I would really regret it if I switched and lost. It's best to stay with your first choice."

Thus, it appeared from this type of comment that people think there will be more negative feelings associated with losing after switching than with losing after sticking. People who switch and are then incorrect may feel especially sorry because they had the winning alternative correctly identified but then switched away from it. People who stick and are incorrect at least know they never had the correct answer in their grasp. Therefore, we designed a study to assess directly the counterfactual view of why people are inclined to stick when they ought to switch.

Students completed a questionnaire which contained the Monty Hall Dilemma, worded similarly to previous word problems. In this study, however, instead of asking people what they would choose, they were assigned on a random basis to a scenario in which they decided to stick and lost *or* one in which they switched and lost. They then rated how they thought they would feel on a series of dimensions, e.g., NOT AT ALL SURPRISED—COMPLETELY SURPRISED. On two of the scales the difference between the conditions was significant. People in the switch and lose condition anticipated they would feel more frustrated and angry than people in the stick and lose condition. These differences imply that counter-factual thinking may be part of the reason why people prefer to stick in the Monty Hall Dilemma after misapprehending the odds at 50:50.

▼

*The Number of Doors*

In her first column on the Monty Hall Dilemma, vos Savant explained her solution in this way:

▼

> Suppose there are a *million* doors, and you pick door number 1. Then the host, who knows what's behind the doors and will always avoid the one with the prize, opens them all except door number 777,777. You'd switch to that door pretty fast, wouldn't you?

So long as the assumptions stated earlier hold, including the *knowledgeable* host, the probability of winning by sticking would then be .000001, and by switching would be .999999. By framing the answer in this way, the implication is that by expanding the number of doors from three to a million, the correct solution of switching becomes more obvious and more compelling.

In a similar vein, William Martin, Chairman of the Department of Nuclear Engineering at the University of Michigan, suggested expanding the problem to one hundred doors with ninety-nine goats and one car. The host would deliberately show ninety-eight goats after the contestant's initial guess. Martin reported using the Monty Hall problem to introduce students to Monte Carlo methods while studying the "Statistical Simulation of Complex Physical Systems."

Writing in the *American Statistician,* Steve Selvin gave a formula devised by David Ferguson, $(N-1)/[N(N-n-1)]$, which yields the probability of winning by switching where N is the number of doors and n is the number of incorrect doors deliberately opened by the host after the contestant makes an initial pick. Even when the host shows only one incorrect door, there is always some advantage in switching, but the advantage gained approaches zero as the number of doors becomes large. On the other hand, if all of the incorrect doors except one are shown, then the advantage of switching increases as the number of doors increases and approaches 1.0 when the number of doors is very large. With only three doors, showing only one incorrect door or all but one incorrect door is, of course, the same thing.

The implication is that if only people would expand the number of alternatives, then they would get it right. There is, however, no proof that expanding the number of doors makes it easier for people to get the problem correct on their initial encounter. In another of our laboratory experiments, we manipulated the number of doors and the number of incorrect doors shown. In our baseline three doors—show one condition, 11 percent (10/87) decided to switch in their first encounter with the Monty Hall Dilemma, compared to 13 percent (6/47) in the five doors—show three condition and 11 percent (5/45) in a seven doors—show five condition. These small differences are not significant. Therefore, we doubt that the initial inclination to stick when one should switch is reduced by making the objective probabilities more compelling than .67 to .33. Since we have not expanded the number of doors in an experiment beyond seven, we cannot extrapolate with confidence beyond that level.

We did find that expanding the number of doors made it easier to learn inductively the solution to the Monty Hall Dilemma. As in our other experiments, subjects in the number of doors study played fifty trials of the dilemma with feedback after each trial and the incentive of a $25 prize for the highest score in each condition. In the last ten trials, 59 percent of the decisions were to switch in the three-door condition, compared to 76 percent in the five doors—show three condition and 87 percent in the seven doors—show five condition. Yet on the criterion of complete mastery, i.e., switching on all of the last ten trials, only 9 percent of the subjects in the baseline three-door condition met this demanding standard, compared to 28 percent in the five-door—show three condition and 44 percent in the seven door—show five condition. Recall that in the latter condition, the probability on each trial of winning by switching is .86 and only .14 of winning by sticking. That is, the advantage of switching is rather pronounced in that condition. Yet after forty trials with feedback, less than half of the subjects in that condition had learned that they should always switch.

▼

*Doors with Unequal Probabilities*

One of the initial assumptions in the Monty Hall Dilemma was that the winning prize be placed behind a door on a random basis with each door having an equal chance of being the winner. This is a good starting point and may be the only reasonable assumption absent evidence to the contrary.

However, suppose the three doors are known to have different probabilities of being the winner. For example, by watching the show over time, a person notices that door 1 is the winner 45 percent of the time, door 2 40 percent of the time, and door 3 only 15 percent of the time. Thus, assumption A is violated, but the remaining assumptions B through F are still valid. Moreover, within these quotas, the placement for a given contestant lacks predictability. The contestant begins by picking door 1, the most likely alternative. The host, taking this into account together with his knowledge of where the car is, opens door 2 to reveal a goat and then allows the contestant to stick or switch.

Should the contestant breathe a sigh of relief since the second most likely alternative has been eliminated, and stick with door 1 which had the highest initial probability of being correct? Or should the contestant switch to what had been the least likely alternative, door 3? One solution is to switch to door 3 since it has an unconditional probability of .55 of being the winner, compared to .45 for the other remaining alternative, door 1. In this view, with a *knowledgeable* host selecting the unchosen door to open, it is as if an "iron curtain" descends around the first pick, so that the probability that had been associated with the incorrect door that is opened transfers entirely to the remaining alternative. So long as the initial probability of door 1 having the car is less than .5, according to this strategy, one should switch.

But wait! For this problem, there is a *better* strategy for the contestant to follow, suggested to me by Craig Anderson. Assuming the contestant knows the probabilities given above and knows the procedure to be followed by the host, the contestant can maximize the likelihood of winning the car by first picking the *least* likely alternative, door 3, and then switching to the remaining alternative after door 1 or door 2 is opened to display a goat. In that way, one can have an unconditional probability of .85 of winning the car by switching. Although we have not tested this version of the dilemma in any of our experiments yet, Anderson's solution is almost certainly counter-intuitive, and we expect it will be extremely rare that someone comes up with it spontaneously.

In fact, this problem with doors that have unequal probabilities turns out to be even more complex. People writing letters to Marilyn about the three doors problem debated whether this was a problem involving conditional probability. But in the case of doors with equal probability, it does not matter whether one uses unconditional or conditional probability since the solution is the same. However, when the doors have unequal probabilities, one can use the additional information of *which door* is opened by the host to condition the probability. Given the choice, conditional probabilities should be used since more information is available and thus, the uncertainty is reduced.

Applying the Bayesian formula for conditional probability to the problem of the doors with probabilities of .45, .40, and .15, respectively, the following probabilities are generated:

| Initial Pick | Door Shown | P Win Stick | P Win Switch |
|:---:|:---:|:---:|:---:|
| 1 | 2 | .60 | .40 |
| 1 | 3 | .36 | .64 |
| 2 | 1 | .57 | .43 |
| 2 | 3 | .31 | .69 |
| 3 | 1 | .16 | .84 |
| 3 | 2 | .14 | .86 |

With these probabilities in hand, it is apparent that if one picks door 1 initially, one should stick if the host opens door 2 but switch if door 3 is opened. Similarly, if one picks door 2, stick if door 1 is opened but switch if door 3 is opened. If one picks door 3 initially, switch regardless of which door is opened by the host. Note, however, that Anderson's solution is *still* the *optimal* one: select the door with the lowest initial probability and then switch!

▼

### Multi-Stage Monty Hall Dilemma

In the three-door Monty Hall Dilemma, there are two stages to the decision, the initial pick followed by the decision to stick with it or switch to the only other remaining alternative after the host has shown an incorrect door. An intriguing extension of the basic Monty Hall Dilemma has been provided by M. Bhaskara Rao of the Department of Statistics at North Dakota State University. He analyzed what happens when the dilemma is expanded beyond two stages. The number of stages can be as many as the number of doors minus one.

Suppose there are four doors, one of which is the winner. The host says:

▼

> You point to one of the doors, and then I will open one of the other doors to show a goat. Then you decide whether to stick with your initial pick or switch to one of the remaining doors. Then I will open another door (other than your current pick) to show a second goat. You will then make your final decision by sticking with the door picked on the previous decision or by switching to the only other remaining door.

Now there are three stages, and the four different strategies can be summarized as follows:

| Stage | 1 | 2 | 3 | Probability of Winning |
|-------|------|--------|--------|------------------------|
|       | Pick | Stick  | Stick  | .250                   |
|       | Pick | Switch | Stick  | .375                   |
|       | Pick | Stick  | Switch | .750                   |
|       | Pick | Switch | Switch | .625                   |

People who accept the correctness of the 2/3 solution in the basic Monty Hall Dilemma might assume that one does best by switching in both Stage 2 and Stage 3. However, as shown here, the counter-intuitive solution to the three-stage Monty Hall Dilemma is to stick in Stage 2 and switch in Stage 3. These remarkable probabilities were published by Rao in the *American Statistician,* and have been verified by Ken Mueller, a student assisting me, using computer simulation. The underlying principle is that in a multi-stage Monty Hall Dilemma, one should stick with one's initial hunch until the very last chance and then switch.

▼

*Why Won't People Switch in the Monty Hall Dilemma?*

Our results imply that for solving two-stage problems, the human brain is not well equipped to process conditional information. Rather than being objective information processors, humans may be wired to choose conservatively to stick with a tentative selection, even when there was no rational basis for that selection, and when it would be perfectly rational to switch to another alternative. Such an inference would be fully warranted only after some cross-cultural checking to see how generalizable our findings are.

Why do people tend to stick in the Monty Hall Dilemma when they ought to switch? The simple answer is that they misapprehend the probabilities of winning by sticking and switching. This begs the question of why people misapprehend the probabilities. One answer is that people ignore, or fail to appreciate fully the importance of the cue of the knowledgeable host. The third question is why do people stick even if they incorrectly assess the probabilities of winning by sticking or switching as .50. Even if the true probability of winning by sticking were .50, there would be no rational basis for sticking. It is on this third question that some interesting social psychological mechanisms come into play.

The counter-factual hypothesis suggests that in deriving their final decision in the Monty Hall Dilemma, people may use the cognitive heuristic of mental simulation. For example, they may ask implicitly, "How will I feel if I switch and lose?" Perhaps affect is greater when the situation involves action rather than inaction, and in this context, switching would represent action and sticking inaction. People may feel worse, and therefore be more likely to recall, when they change an answer on a multiple choice exam and it turns out to be incorrect than when they stick with a doubtful answer and it turns out to be incorrect. In the Monty

Hall Dilemma, we found that people thought they would be more frustrated and angry if they switched and lost than if they stuck with their initial selection and then lost.

Another basis for hypothesizing a tendency to stick with an initial hunch in the Monty Hall Dilemma is to see it as an example of the more general phenomenon of *belief perseverance*. Lee Ross and Mark Lepper of Stanford University have shown that once people have formed a belief, it is difficult to disabuse them of it. People are led to believe, for example, that they are very good *or* very poor at identifying real suicide notes. They continue to believe this even after a concerted effort is made to convince them that they had been randomly assigned to receive false feedback of one kind or the other. However, perseverance studies have dealt with beliefs about the world, e.g., whether the death penalty deters murder or whether high risk-takers make good firefighters, beliefs that involve the marshalling of evidence and reasoned argument. By contrast, the Monty Hall Dilemma involves nothing more than a hunch or guess. Also, the Monty Hall Dilemma involves perseverance in behavior rather than perseverance in belief.

Ellen Langer of Yale University developed the concept of *illusion of control* which has many applications to human cognition, including perhaps the Monty Hall Dilemma. People could experience an illusion of control when making the initial guess, and this may make them reluctant to give up that alternative later in the absence of a compelling reason. In this view, switching might be more likely if the initial selection is made by another person. We tested this recently by randomly assigning people to play the three-door game for fifty trials in one of two conditions. In the standard baseline condition, in which people made both the initial and the final choice, only 9 percent switched on the first trial. In the new condition, in which another person made the initial selection and the subject made only the final decision of whether to stick or switch, 38 percent witched on trial 1. This difference between conditions is significant and continued through twenty trials. Curiously, on trials twenty-one to fifty, people in the two conditions switched at about the same frequency. Both conditions showed the plateau function, and in the final ten trials, there was 55 percent switching in the baseline condition and 56 percent in the new condition.

A fourth question concerns why people have such a difficult time learning inductively to switch across fifty trials. People in the baseline condition begin at a very low level of switching, and then gradually increase their switching to a plateau between 50 and 60 percent. After viewing the leveling off at slightly more than 50 percent switching in the Monty Hall Dilemma and the flat trajectory in the Russian Roulette Dilemma, we wonder whether allowing the experiment to continue for one hundred or one thousand trials would have altered the results.

One impediment to learning is that most people do not try playing one strategy consistently. Rather, they play a mixed strategy of sticking on some trials and switching on others. If people could be led to try one strategy consistently over a number of trials, they might be more likely to catch on to the solution of always switching in the Monty Hall Dilemma. Our student subjects may have had a misplaced focus. Told in advance there would be fifty trials, many subjects seemed to be looking for non-random sequences in the correct answer across trials rather than focusing on the basic structure of the situation in a given trial. It probably never occurred to most subjects that the correct solution might be to *never* switch or to *always* switch. Playing a mixed strategy probably makes it more difficult to discern the correct solution, despite the 2:1 odds for winning by switching. Many subjects may have found themselves winning on about half the trials, or slightly better, and may have been satisfied or felt they could do no better. Herbert Simon's concept of *satisficing,* in which people settle for a satisfactory level of winning rather than search for an ideal or optimal solution, may help to understand the behavior of our subjects.

The solution to the Russian Roulette Dilemma seems to be less counter-intuitive than the Monty Hall Dilemma. Yet, most subjects who played the Roulette Dilemma for fifty trials did not come close to getting it completely correct. Some of them may think correctly that the odds of winning by sticking are 2:1 in the Roulette Dilemma, but conclude incorrectly that they should lean toward sticking and stick on about 2/3 of the trials. This sort of strategy is called *probability matching*. In fact, the odds of winning by sticking are 2:1 in the Roulette Dilemma, and therefore one should stick on every trial. By doing so, one can expect to win on about 67 percent of the trials while a probability matching strategy yields a win on about 56 percent [ $(67 \times .67) + (33 \times .33)$ ] of the trials.

If we stretch things, what began as an intriguing counter-intuitive problem in probability may provide a model for studying rational and irrational tendencies in two-stage decisions. Our results may be relevant to the psychological processes associated with decision making. Leon Festinger's theory of cognitive dissonance held that the interesting psychological processes of subjective distortion, selective exposure and rationalization occur only after a final and irrevocable decision has been made. It may be that this view is partially incorrect in that there may be some intriguing processes that occur after a preliminary and entirely revocable decision has been made. Even when people have no good reason for their initial choice, having acted upon it, they may become psychologically committed to it.

Of course, it is fair to ask whether our experimental results have anything to do with how people make decisions in real life. On the one hand, we think that two-stage decisions, in which people make a tentative decision, gain additional information, and then make a final commitment, are very common, we might even say ubiquitous. A couple becomes engaged and later is married. A military commander makes a preliminary decision to make an attack at a particular place, e.g. Normandy Beach, and then continues to monitor information about relevant alternatives before deciding finally when and where to launch the attack. A university president makes a tentative decision to hire a particular candidate to be basketball coach, but mulls it over in her mind while continuing to receive additional information, pro and con, before making a public announcement. If people really have an irrational tendency to stick with an initial decision when they ought to switch, this would obviously be an important finding.

On the other hand, the Monty Hall Dilemma may be difficult for people to comprehend and solve because it is so uniquely contrived. Aside from the game show context, it is not easy to come up with another situation in which people receive valid information about unchosen alternatives from a *knowledgeable source* after a tentative decision but prior to a final commitment. But scientific experiments, whether in chemistry or psychology, often create situations that would never occur under natural circumstances. The Monty Hall Dilemma may be a case in point, and its contrived nature does not mean that we cannot learn something from the data that have been generated.

Thus, if the Monty Hall Dilemma (where one should switch), is considered in conjunction with the Roulette Dilemma (where one should stick), we have a framework for analyzing the two-stage decision process.

Much of the research on two-stage decisions, e.g., that dealing with the intention-behavior relationship, does little more than identify the switchers and the standpatters and compare them. The researchers often can say nothing about whether it is rational to switch from intention to behavior. Our results imply that people stick when they should stick, but they also tend to stick when they should switch. This may indeed say something important about how people make two-stage decisions.

# Index

# About the Author

Marilyn vos Savant was born in St. Louis, Missouri, the daughter of Marina vos Savant and Joseph Mach. She is married to Robert Jarvik, M.D., the inventor of the Jarvik-7 artificial heart. They live in Manhattan. Marilyn vos Savant was listed in the *Guinness Book of World Records* for five years under "Highest IQ" for both childhood and adult scores, and she has now been inducted into the Guinness Hall of Fame. She is a writer, lecturer, and spends additional time assisting her husband in the artificial-heart program. She is a member of the Board of Directors of the National Council on Economic Education and a member of the National Advisory Board of the National Association for Gifted Children. Her special interests are politics and leadership; quality education and thinking; humanitarian medicine and research. She describes herself as an "independent" with regard to politics and religion, and only an "armchair" feminist.

Marilyn vos Savant writes the "Ask Marilyn" question-and-answer column for *Parade,* the Sunday magazine for more than 340 newspapers nationwide, with a circulation of 37 million and a readership of 81 million, the largest in the world. Past book publications include: *The Power of Logical Thinking,* published in hardcover; *More Marilyn,* published in hardcover and trade paperback; *"I've Forgotten Everything I Learned in School!",* published in hardcover and trade paperback; *The World's Most Famous Math Problem: The Proof of Fermat's Last Theorem and Other Mathematical Mysteries,* published in trade paperback; and *Ask Marilyn,* published in hardcover, trade paperback, and mass market; all by St. Martin's Press.